Advances in Industrial Control

For further volumes:
www.springer.com/series/1412

Igor Boiko

Non-parametric Tuning of PID Controllers

A Modified Relay-Feedback-Test Approach

 Springer

Igor Boiko
Electrical Engineering Department
The Petroleum Institute
Abu Dhabi
United Arab Emirates

Additional material to this book can be downloaded from http://extras.springer.com.

ISSN 1430-9491   ISSN 2193-1577 (electronic)
Advances in Industrial Control
ISBN 978-1-4471-6046-5   ISBN 978-1-4471-4465-6 (eBook)
DOI 10.1007/978-1-4471-4465-6
Springer London Heidelberg New York Dordrecht

Printed on acid-free paper

Springer is part of Springer Science+Business Media (www.springer.com)

To my family

Series Editors' Foreword

The series *Advances in Industrial Control* aims to report and encourage technology transfer in control engineering. The rapid development of control technology has an impact on all areas of the control discipline. New theory, new controllers, actuators, sensors, new industrial processes, computer methods, new applications, new philosophies..., new challenges. Much of this development work resides in industrial reports, feasibility study papers and the reports of advanced collaborative projects. The series offers an opportunity for researchers to present an extended exposition of such new work in all aspects of industrial control for wider and rapid dissemination.

The proportional-integral-derivative (PID) controller continues to be an important controller for industrial applications. It is mainly found at the lower level of the control hierarchy, at the actuator-sensor level, and configured into multiloop structures (for example, cascade loops). Groups of process units may be provided with optimised set points using advanced techniques like model predictive control, but the local process loops are likely to be PID controlled. The ease of understanding of PID controller tuning by maintenance and operational staff and the widespread availability of PID functionality in distributed-controlsystem/SCADA software and programmable logic controllers (PLCs) are important factors that support the continued industrial use of PID control.

Despite this favourable situation for PID control, the large numbers of PID controllers in industrial-scale process plant, the economic pressure to contain or reduce maintenance costs and the enabling advances in industrial process computer technology led to interest in creating techniques for the automatic tuning (termed autotuning in the 1980s) of PID control loops. In this control-tuning field, the work of Ziegler and Nichols in the 1940s initiated a procedural framework for the rulebased tuning of PID loops. In the 1980s, the work of Åström and Hägglund involving a relay-test-based procedure, gave the field a new impetus and enabled autonomous tuning to become an online reality.

The methods of PID controller tuning involve two steps:

Step 1 A Measurement Experiment—in which some chosen characteristics of the process are measured. These may be measurements to allow a process model to be identified; they may be measurements of a characteristic that can be used directly in the computation of controller parameters.

Step 2 Controller Tuning Computation—in which control loop performance requirements are specified, the structure of the controller is decided and the controller parameters are computed using the data coming from Step 1.

Automated PID controller tuning seeks theory, algorithms and implementations that minimise the use of expert input (from, for example, technical, maintenance, or operational staff) in the tuning process and allows the autonomous tuning of PID controllers using computer software and hardware installed at the plant. These techniques comprise two categories:

Parametric tuning methods, where the measurements are used to identify a process model and the controller tuning uses the identified process model data.

Non-parametric tuning methods, where the measurements are of characteristics or parameters that are used directly in formulas for the computation of the controller parameters.

The "continuous cycling" procedure and its associated tuning rules devised by Ziegler and Nichols, and the relay test procedure and the associated phase margin tuning rules are examples of the non-parametric family of tuning methods. These two contributions have been hugely influential in this particular field of industrial control practice, and they continue to inspire current day research and development.

Monograph contributions to this field in the *Advances in Industrial Control* series cover both PID control and the Automatic PID controller tuning field and those in the PID control field include:

2012 R. Vilanova and A. Visioli (Eds.): *PID Control in the Third Millennium: Lessons Learned and New Appoaches*, ISBN 978-1-4471-2424-5;
2010 A. Visioli and Q.-C. Zhong, *Control of Integral Processes with Dead Time*, ISBN 978-0-85729-069-4;
2006 A. Visioli, *Practical PID Control*, ISBN 978-1-84628-585-1;
1999 A. Datta, M.T. Ho and S.P. Bhattacharyya, *Structure and Synthesis of PID Controllers*, ISBN 978-1-85233-614-1; and
1999 K.K. Tan, Q.-G. Wang and C.C. Hang with T.J. Hägglund, Advances in PID Control, ISBN 978-1-85233-138-2;

whereas, monograph contributions to the automatic PID control tuning field include:

2011 T. Liu and F. Gao, *Industrial Process Identification and Control Design*, ISBN 978-0-85729-976-5; and
1999 C.C. Yu, *Autotuning of PID Controllers: Relay Feedback Approach*, ISBN 978-3-540-76250-8.

To this activity, it is a pleasure to add this monograph, *Non-Parametric Tuning of PID Controllers: Modified Relay Feedback Test Approach*, by Igor Boiko. Despite an extensive literature on the theory and application of the relay test procedure Igor Boiko brings new insights to the field based on his industrial experience by proposing that the tuning and the experimental procedure should be matched to the features of specific categories of control loops found in the process industries. In the monograph analyses of the PID controller requirements for flow, level, pressure, and temperature loops are presented. Controller performance is based on gain and phase-margin criteria conjoined with time integral (IAE, ITAE, ISE, ITSE) optimality. This novel approach will be especially interesting to the industrial control practitioner.

Later chapters in the monograph describe an investigation into the effects of process nonlinearity on the tuning results for flow loops, and present more theoretical details of the modified relay test procedure reported in earlier chapters. The closing chapter of the monograph examines some of the practical issues involved in performing online automatic tuning methods and describes software for the tuning task.

The mix of practical expertise and the theoretical elaboration presented by Igor Boiko in this *Advances in Industrial Control* monograph will attract a wide readership from the control community and the process industries and continues the series tradition of publishing exemplary research and development in the PID control paradigm.

Industrial Control Centre   M.J. Grimble
Glasgow, Scotland, UK   M.A. Johnson

Preface

The subject of this book is the method of PID controller tuning based on the continuous cycling principle. Relay feedback for tuning was proposed by Åström and Hägglund in the 1980s and has been completed since then by numerous modifications aimed at enhancing some features of the original method. The majority of these modifications concern parametric methods of tuning that are based on the identification of certain underlying process models. The method presented herein is non-parametric. It features a *holistic* approach to test and tuning, or *coordinated test and tuning*, in which the test parameters are selected not arbitrarily or a priori but together with the tuning rule to be applied. As a result, this method provides exact values of a specified gain or phase margin and does not require any iterative procedure. Another novel feature is the introduction of *process-specific optimal tuning rules* in the non-parametric setup. This allows an engineer to use the flow loop-optimised tuning rules for flow loop tuning, level loop-optimised tuning rules for level loop tuning and so on, and obtain in most cases a better result than generic tuning rules would yield.

We hope readers might also find the presented approach to obtaining optimal tuning rules an interesting one. It involves a nontraditional solution of the optimisation problem. We also believe that the recently developed *dynamic harmonic balance* principle, which is presented in this book, may attract reader interest as well.

A person studying the subject of automatic control might benefit from this book by learning how linear and nonlinear control theory are brought together to solve a very important practical control problem—optimal tuning of PID controllers. A practising control engineer might gain new insights into PID controller tuning. The presented method is simple in realisation and efficient in terms of practical results. The reader can use the provided MATLAB code to incorporate the various components of the presented theory and tuning method.

When deriving tuning rules for particular types of industrial processes, focus is given to the most common process industry applications: flow, temperature, pressure and level control loops. It is assumed in most situations that the process is manipulated through control valves (or dampers), and more specifically through pneumatically actuated control valves. In various industries different features of loop tuners are important. In this book we assume that tuning could be carried out on a live process (i.e. during plant operation), so that such features as minimum disturbance to the process and short time required for tuning are of great importance.

Chapter 1 is introductory. It traces the history of tuning—and in particular, non-parametric tuning. Differences between parametric and non-parametric approaches are discussed. The issue of the selection of proper complexity of the process model is illustrated by an example of curve fitting.

Chapter 2 covers the methods of non-parametric tuning of PID controllers. The closed-loop Ziegler–Nichols method and Åström–Hägglund's relay feedback test are reviewed. It is shown that they allow one to generate test oscillations at the point corresponding to −180° of the phase characteristic of the process. However, it is also shown that generation of oscillations in the third quadrant would be beneficial. A number of non-parametric tuning methods based on various modifications of the relay feedback test (RFT), which are capable of producing test oscillations in the third quadrant, are reviewed.

In Chap. 3, the modified relay feedback test (MRFT) is introduced as a further logical development of the closed-loop Ziegler–Nichols method and Åström–Hägglund's relay feedback test. Methods and criteria of tuning are presented in general and for every considered typical process: flow, level, pressure and temperature processes. With the criterion of optimisation selected and the mechanism of disturbance generation analysed for each of the considered processes, the optimal tuning rules are obtained by solving the *optimisation problem on the domain of parameters* characterising the situational aspects of the implied process model.

Chapter 4 illustrates possible ways to improve the accuracy of the tuning rules via an example of the flow loop. It is shown that the precise model of the flow process is nonlinear even if the installed characteristic of the valve is linear. This nonlinearity is revealed as an apparent time constant of the actuator dynamics, which depends on both the amplitude of the relay test and the selection of the operating point. It is shown that we benefit by using the precise nonlinear model of the flow loop when finding optimal PID tuning rules.

Chapter 5 covers the exact model of oscillations in the system arising from the modified relay feedback test. The exact model can be used for parametric tuning that includes identification of the process model parameters. The development here is based on the *locus of a perturbed relay system* (LPRS) method. We present the LPRS for the conventional relay feedback system, the MRFT and the test containing the two-relay controller.

Chapter 6 provides the model of transient oscillatory motions in a system using the modified relay feedback test. The treatment is based on the concept of the *dynamic harmonic balance*.

Chapter 7 covers the practical implementation of this book's tuning method in software for distributed control systems. The most critical issues encountered in the implementation are covered. Industrial software used in the process industry is described.

Some chapters of the book can be read independently of others. For example, practising engineers may be more interested in Chaps. 1, 2, 3 and 7. They can omit other chapters without sacrificing an understanding of the main ideas of non-parametric tuning and the modified relay feedback test in particular. At the same time researchers may find the material of Chaps. 5 and 6 interesting from the perspective of finding exact models of the oscillations and analysis of transient oscillatory modes, respectively.

I express my heartfelt gratitude to Prof. M. Johnson for his careful reading of the manuscript and numerous comments, which allowed me to significantly improve the book; to Prof. M. Grimble for his initiation of this project and encouragement of my work on the book; to my co-workers and colleagues at Syncrude Canada: D. Brown, A. Ernyes, W. Oli, and E. Tamayo; to my former MSc student S. Sayedain, who did simulations presented in Chap. 4; to A. Breslavskaya, who produced a significant share of the artwork and LaTeX typesetting for the book; and to K. McKenzie for correcting and improving the quality of language of the manuscript. I am also grateful to my family for their patience and the personal sacrifices which they have given to this my work. Without their support this undertaking would not have been possible.

I also gratefully acknowledge the support of RIFP Project No. 12310 of the Petroleum Institute, Abu Dhabi.

Abu Dhabi   Igor Boiko

Acronyms

DCS	Distributed control system
DF	Describing function (method)
DHB	Dynamic harmonic balance
DSOPDT	Damped second-order plus dead time (model/dynamics)
FOPDT	First-order plus dead time (model/dynamics)
HB	Harmonic balance
HMI	Human-machine interface
IAE	Integral absolute error
ISE	Integral square error
ITAE	Integral time absolute error
ITSE	Integral time square error
I/P	Electric current-to-pressure (transducer)
LPRS	Locus of a perturbed relay system
MRFT	Modified relay feedback test
PID	Proportional-integral-derivative (controller or algorithm)
PLC	Programmable logic controller
RFT	Relay feedback test
SISO	Single-input-single-output (system)
SM	Sliding mode
SOPDT	Second-order plus dead time (model/dynamics)

Contents

Chapter 1
Introduction

1.1 Historical Overview

Process automation technology has taken a huge leap forward since the 1980s. The processing power of controllers has grown dramatically and influenced the control strategies used in applications. The complexity of control algorithms has grown with computing power. Multi-loop and multi-variable controllers are used now in applications where only single loops were once used. Yet proportional-integral-derivative (PID) control remains the main building block of these control strategies.

Significant progress can also be seen in the area of PID controller tuning. Loop tuning previously was a very complicated, hands-on activity; it can now be routinely done by specialised software. However, engineers still need to understand loop tuning principles, though they now can carry out this activity very efficiently. Yet, despite the sophistication of many modern tuning methods, the ideas can be traced back to the early works on PID control and tuning.

By the mid 1940s of the last century the PID controller had become widely used in the industry. Theoretical foundations of this type of control were laid by N. Minorsky [62] and other researchers. However, a general approach to tuning PID controllers had not been developed yet. A few tuning techniques were available, which showed satisfactory results on some processes but were totally useless when applied to others. Both theoreticians and practitioners had been looking into the problem of tuning of PID controllers but their efforts had not produced the theory of PID controller tuning yet. According to A. MacFarlane [56], "By the late 1930's there were thus two separate but well-developed methods of attacking the analysis of feedback system behaviour. (1) The "time-response approach" ... and (2) the "frequency-response approach"". The advent of a solid approach to PID controller tuning as well as the introduction of non-parametric methods of controller tuning are due to the research of John G. Ziegler and Nathaniel B. Nichols presented in their seminal work [88]. In that 1942 publication, Ziegler and Nichols proposed two methods: the open-loop tuning method, which included the step test on the process, and the closed-loop tuning method, which involved the test featuring a continuous oscillation. Generally, these were in line with the two trends of control theory of that time.

I. Boiko, *Non-parametric Tuning of PID Controllers*, Advances in Industrial Control, DOI 10.1007/978-1-4471-4465-6_1, © Springer-Verlag London 2013

The former method was a time-domain method and bore the feature of both: parametric and non-parametric loop tuning, whereas the latter was a frequency-domain, purely non-parametric method. In [14], George Blickley noted: "There were about 15 other mathematical routines that could be used instead of the Ziegler–Nichols method, and each was tried and championed—only to succumb to the simplicity and ease of use of Ziegler–Nichols tuning. ... The control industry practically snubbed the Ziegler–Nichols method when it was introduced to the ASME, but it soon gained wide acceptance over intuitive and flawed methods used at the time. If there is ever a museum built to honour PID, statues of John Ziegler and Nat Nichols should be at the entrance." Despite that fact that the Ziegler–Nichols tuning method is now rarely used in its original formulation, since it often results in an oscillatory response, its influence on the development of other tuning methods is enormous.

Since 1942 a large number of tuning rules and algorithms have been developed and presented in the literature. Many of these can be found in [66]. However, new algorithms and rules are still being developed, as new ideas are generated and new technologies capable of accommodating new tuning methods are developed. What can be emphasised is that most tuning methods rely on tests carried out over the process, while there are some others which do not force tests but utilise observation of the process fluctuations caused by external disturbances. The second category of tuning methods constitutes the class of non-invasive tuning methods. These methods are safer because they do not involve any disturbances to the process. Yet the magnitude of the disturbances acting on the process is not always sufficient for reliable identification of process dynamics, This is the source of the lower accuracy of these methods, which are still in development and more seldom used in practice than the invasive methods. Therefore, methods which involve tests over the process nowadays provide the main biggest category of tuning algorithms and rules.

The two main types of tests used for process dynamics identification and controller tuning are the step test and the test involving continuous oscillation of the process variable due to the feedback. Other types of test signals are used too, the following most frequently: the ramp signal, a series of randomly generated pulses or pulses generated according to a certain law [43], single or multi-frequency harmonic excitation of the process [3].

Despite some specific advantages of each particular method, because its development was motivated by the desire to solve a particular problem, we consider in this book only those methods which involve generation of a continuous oscillation through the feedback. The author believes that this is a group of methods allowing one to most fully use the test itself and obtain maximum information from a single test. This is owing to the remarkable property of all the tests from this category: the oscillations generated by the test produce the frequency which is most informative and important to the controller tuning. In comparison, for example, with methods that utilise external harmonic excitation, one does not have to solve the problem of selecting the frequency of the external excitation, because this frequency is generated automatically through the use of the feedback, and the generated frequency is optimal or close to optimal in terms of the accuracy of the subsequent tuning.

The development of the tests, which excite a continuous oscillation by utilising the feedback signal, has evolved through several stages. The first and most celebrated method was proposed by Ziegler and Nichols, as noted above. The continuous cycling test was adopted by a number of researchers afterwards, who proposed different tuning rules still based on the measurement of the ultimate gain and ultimate frequency. Such popular tuning rules as those by Cohen–Coon [30] and Tyreus–Luyben [54] are examples. Integral performance criteria-based formulas were developed in [72]. Such tuning formulas as those of [52] and [63], although not directly applicable to non-parametric tuning, also influenced this area by providing tuning criteria. More tuning rules can be found in [66].

The next significant step in the development of the continuous oscillation methods was made in the work of Åström and Hägglund [5], who proposed the *relay feedback test*. The test itself is based on the relay feedback principle, in which the term "relay" finds its origin in electrical applications where on-off control has been used for a long time. The nonlinear function that describes the nonlinear phenomenon typical of electrical relays is also named by the term "relay"; it comprises a number of discontinuous nonlinearities. When applied to this type of control systems "relay" is now mostly associated not with applications but with the kind of nonlinearities found in system models. This test was a significant improvement of the closed-loop Ziegler–Nichol's test because it provided virtually the same measurement results but offered an opportunity for automatic loop tuning. In the relay feedback test, instead of a trial-and-error procedure, only one experimental test is required to excite a sustained oscillation and obtain the values of the ultimate gain and ultimate frequency. This test has attracted a lot of attention from the researchers and practitioners. A number of modifications of this test were proposed in publications [21, 41, 47, 85–87] and others.

In its development, it became apparent that the relay feedback test did not provide the oscillation frequency optimal for the identification of process parameters, and the model of the oscillations based on the describing function method gave only an approximate solution of the identification problem. Many attempts were made to improve the method in both aspects. Attempts to improve the test through exciting oscillations at the frequency corresponding not necessarily to $-180°$ but to other specified angles as well were made in the following works. In [78] it was proposed that a delay and iterative process of the update of delay time should be used to excite oscillations in the third quadrant of the process frequency response. Use of an additional integrator was proposed in [38], use of a differentiator in [29]. A phase-lock loop to excite oscillations at the frequency corresponding to a specified phase lag was suggested in publications [33], [35] and [36]. These methods are described in this book. However, the presented material focuses primarily on the author's modified relay feedback test, proposed in [19] and presented in more detail in [27]. This method not only includes identification at the specified phase lag but provides a complete approach to *test-and-tuning* as a holistic process in which the test parameters are determined in connection with the selected tuning rules. This approach allows one to provide the specified gain or phase margins in the loop being tuned exactly (within the framework of the describing function method assumptions). This is further referred to in the book as *coordinated test and tuning*.

Fig. 1.1 Mappings of
parametric tuning

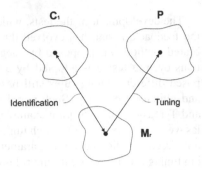

1.2 Parametric and Non-parametric Tuning

There are two groups of loop tuning methods: parametric and non-parametric. Every tuning involves two steps: tests or experiments undertaken over the plant (process) to measure some characteristics of this process; and after that, controller tuning is done on the basis of the measurements obtained during the tests. The organisation of the steps defines whether the method is parametric or non-parametric. In parametric tuning, a model of the process is involved and the above-mentioned tuning step includes identification of the parameters of this model and tuning itself, which is based on the available methods of control system design. Parametric tuning can be illustrated as the following two mappings (Fig. 1.1): $C_1 \rightarrow M_r$ and $M_r \rightarrow P$, where C_1 is the space of measurable characteristics of the process, M_r is the space of parameters of the *underlying* model, and P is the space of the controller parameters. Therefore, parametric tuning can be described in terms of the considered space mappings as $T_P : C_1 \rightarrow M_r \rightarrow P$. Yet it is clear that the mapping $C_1 \rightarrow M_r$ is not easy to obtain. Rather what can be easily modelled and simulated is the inverse mapping $M_r \rightarrow C_1$. The mapping $C_1 \rightarrow M_r$ can be obtained through various techniques of identification of the parameters of the model (coordinates of a point in M_r).

Besides the problem of designing computationally efficient and practically feasible identification algorithms, which includes the selection of appropriate tests over the process, there is one more fundamental problem in all parametric methods of tuning. It is the problem of the selection of an appropriate *underlying* process model. In Fig. 1.1 the space of the parameters is denoted by M_r, where r stands for *reduced*. This is because the true model is actually unknown. We can approach this ideal, or true, model more accurately by considering smaller details of the process such as increasing the order, including nonlinearities, etc. However, we can never obtain an absolutely precise model because some even smaller details of the process would still remain unseen. Furthermore, to ensure that the mapping $C_1 \rightarrow M_r$ can be obtained we have to increase the dimension of the space C_1, so that it not be smaller than the dimension of the space M_r. Or, in simple terms, we have to increase the number of measurable characteristics of the process or the number of tests, something that may be impractical. Therefore, there is the problem of the discrepancy between the true model (which is unknown) of the process and the selected *underlying* model. The ability to formulate a precise model to a large extent determines the accuracy and robustness of a particular method of tuning.

Fig. 1.2 Curve fitting into data set

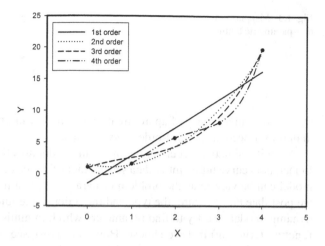

Underlying models for parametric tuning may be either based on fundamental laws of physics and chemistry (first principles) or they may be empirical. In either case the model structure is postulated and the problem of building the underlying process model is formulated as the problem of identifying the parameters of this postulated model. In empirical modelling, the most common approach prefers the simplest model structures, provided this it is physically reasonable to do so [51].

The same approach—one seeking simplicity of the model structure—can also be applied to models based on first principles due to the problem of limited *accuracy of identification subject to measurement errors*. This issue can be illustrated by an example of approximation of data set obtained with errors of measurement.

We assume that the data set is obtained experimentally with certain limited accuracy. Let the experimental data represent the true functional dependence given as $y = x^2 + 1$, and the measurements taken at the points $X = [0; 1; 2; 3; 4]$ as containing some errors within the 20 % error distribution and given as follows: $Y = [1.2; 1.7; 5.8; 8.2; 20.0]$. We will try to approximate the experimental data with polynomials of first, second, third and fourth order minimising the sum of squares of the errors between the experimental data and the value of the polynomial approximation in respective points. The results obtained using the MATLAB function *polyfit* are as follows. The first-order polynomial approximation (linear regression) is $P_1(x) = 4.41x - 1.44$, the polynomials of orders from second to fourth are as follows $P_2(x) = 1.493x^2 - 1.561x + 1.546$, $P_3(x) = 1.483x^3 - 1.407x^2 + 2.595 + 0.966$, $P_4(x) = 0.683x^4 - 4.983x^3 + 11.967x^2 - 7.167x + 1.2$. The measured data and the plots for the approximating polynomials are presented in Fig. 1.2. One can see that increasing the order of the approximating polynomial results in improving the accuracy of approximation only in the points having the same argument values as the measured points and does not necessarily enhance the accuracy of approximation in all other points. Keeping in mind the original function $y = x^2 + 1$ and the fact that the measured values just represent this function, and the actual accuracy of interest is the accuracy of the representation of the whole function $y(x)$, one can see

Fig. 1.3 Mappings of
non-parametric tuning

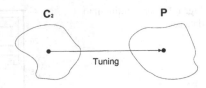

that the overall accuracy of approximation for the fourth-order polynomial is worse
than for the second- or third-order polynomials.

A similar situation occurs in the problem of identification of model parameters.
In fact, both curve fitting into a data set and identification of parameters of a dynamic
model can be viewed as the problem of parameter identification. In both problems
we postulate the structure: the type and the order of the function or the order of the
dynamic model, and try to find parameters which minimise the value of some cost
function (criterion) that we choose. However, as we see from Fig. 1.2, if the data
are obtained with certain errors, an attempt to increase accuracy by increasing the
order of the approximating model may actually result in the deterioration of accuracy. This can be easily seen in the case of curve fitting but it is not obvious in the
case of parametric identification of the dynamic models. Therefore, this issue must
be considered when one selects an underlying model of the process for parametric
tuning.

The non-parametric tuning methods do not rely on parameter identification of
an underlying model. Instead, the controller parameters are computed immediately
from the results of measurements of process characteristics. The whole process can
be represented as a mapping of the space C_2 of measurable characteristics into the
space P of controller parameters: $T_{NP} : C_2 \to P$ (see Fig. 1.3). It can be emphasised that the space of measurable characteristics in parametric tuning is not the
same used in non-parametric tuning. That is why they are denoted as C_1 and C_2 in
Figs. 1.1 and 1.3, respectively. Selection of C_1 and C_2 should meet the specific requirements of the tuning type. Usually, besides satisfaction of some specific requirements coming from the process over which the tests are carried out, the selection of
characteristics C_1 should be done in a way that provides the highest resolution of
model parameter identification. In other words, the sensitivity of the characteristics
to the parameters must be high enough to ensure precise identification of the model
parameters. For example, for identification of small time constants in the process
model one cannot measure the process response to an harmonic signal of low frequency because this response would have low sensitivity to small time constants.
Test signals must be sufficiently fast to provide identification of small time constants. However, this might not always be possible either, because the response may
be masked by the effect of larger time constants.

In the case of non-parametric tuning, the problem of resolution at parameter identification does not exist in a direct sense. However, there is a problem concerning
the selection of measurable characteristics, since they must correlate with characteristics used in closed-loop system performance measures. This rule can be illustrated by the following example. If we design a proportional controller using the
Ziegler–Nichols closed-loop test it would be easy to provide closed-loop systems

with a required *gain margin*. Once we determine the ultimate gain we can easily provide the required gain margin by setting the value of the proportional gain of the controller equal to the quotient of ultimate gain and gain margin. In this case the measurable characteristic being the ultimate gain precisely correlates with the characteristic used for the assessment of system performance (gain margin), which allows one to design a technique that ensures the desired performance. The perfect correlation of these two characteristics means in the considered case that they both depend on the same parameter and not on other parameters (this is the sole parameter of the controller, a situation that does not usually happen).

Therefore, the requirements to the models used in parametric and non-parametric tuning can be summed up as follows. The structure of the *underlying model* used in *parametric* tuning must be simple enough (low-order) to avoid the problem of inaccurate measurements; however, the model should be physically reasonable. The *implied model* of the process used in *non-parametric tuning* can be of arbitrarily high order and complexity because the number of measurements in the test remain low in this case, too (two measurements, for example). The main requirement for this model is that it should provide a test response maximally resembling the response of the actual process. In this case the optimisation of the tuning rules, for which sole purpose the model is used, will give the best results.

1.3 Conclusions

Chapter 1 is introductory. It traces the history of parametric and non-parametric tuning back to its origin. Differences between parametric and non-parametric approaches are discussed. The issue of the selection of proper complexity of the process model is illustrated by an example of curve fitting to experimental data. It is also emphasised that this problem pertains only to the parametric method of tuning, which uses an underlying model of the process. Non-parametric methods of tuning still can use some models at the stage of derivation of optimal tuning rules, but these models (named "implied models") do not suffer the complexity issue. In fact, the opposite is true: the use of more complex and precise models allows one to produce more efficient process-specific tuning rules. The tuning process is totally non-parametric, though.

Chapter 2
Non-parametric Tuning of PID Controllers

As pointed out in the Introduction, there are two approaches to tuning controllers: parametric and non-parametric. Non-parametric methods of tuning based on the continuous cycling principle are considered in this chapter. We start with the Ziegler–Nichols closed-loop tuning method and progress to the available methods of tuning, which involve the possibility of excitation of test oscillations at frequencies corresponding to phase lags of the process other than $-180°$. The necessity of such functionality is supported by examples provided.

2.1 PID Control

Proportional-integral-derivative (PID) control is the main control of the process industry. Research shows that the share of PID controllers in a typical overall plant control system is about 97 % (see [75]). The author's personal observations arising from his work in oil, gas and power plants unequivocally confirm this to be the case. Moreover, if controllers such as ratio controllers (because a ratio controller is not a feedback controller) and ON-OFF controllers are disregarded in this count then the presented figure may be even higher. We also mention that nearly all controllers for "elementary" processes (simple loops like flow, level, pressure and temperature not having much interaction with other loops) are in fact PID controllers. So even if there is a model-predictive controller in the system, for example, this controller produces set points to flow controllers, and flow controllers are implemented through PID algorithms. Therefore, PID control is utilised within the model-predictive control framework, too.

There are a number of features which count towards the advantages of PID control and many reasons why it is so popular. First of all it is a very simple algorithm to implement in the modern programmable logic controllers (PLC) and the distributed control systems (DCS). In fact, many PLCs have a built-in PID control algorithm, so that programming becomes simple. The same feature is seen in the DCS even to a greater degree. Another advantage of PID controllers supplied with a modern DCS

I. Boiko, *Non-parametric Tuning of PID Controllers*, Advances in Industrial Control, DOI 10.1007/978-1-4471-4465-6_2, © Springer-Verlag London 2013

is the variety of features empowering the PID control. Modern DCS have a rich variety of different PID controllers, which are provided with the features of switchable modes of operation, various equations that allow for applying the PID components to either error or process variable, nonlinear gains, the possibility of backtracking and back-initialisation, and several other. Some PID controllers have auto-tuners or they can easily interact with external loop tuning software. The performance of PID control can also be enhanced by introducing lead-lag and other compensators in the loop. Such systems as 2-input-2-output systems can normally be handled well by PID controllers with additional feed-forward signals, too.

There is vast body of literature on PID control. The fundamentals of PID control can be found in almost every book on process control. Some examples of these are (in chronological order) by Shinskey [76], Ogunnaike and Ray [68], Luyben and Luyben [54], Marlin [57], Corripio [32], Bequette [13], Seborg et al. [75], Corriou [31], Ellis [37], Altmann [4]. Detailed presentation of PID control is given in the books by Åström and Hägglund [6] and [7], Tan et al. [79], Visioli [82] and Johnson and Moradi [44].

PID control is a three-component control, the components being the proportional, the integral and the derivative. In the so-called expanded, or noninteracting form, the equations of the PID controller can be written in the Laplace domain as follows:

$$W_c(s) = K_c + \frac{K_i}{s} + K_d s. \qquad (2.1)$$

In the expanded equation of the PID control, all the components are mutually independent and a change of any gain would change this component alone. The proportional, the integral, and the derivative gains K_c, K_i, and K_d, respectively, are referred to as tuning parameters. However, this equation is rarely used in practice and most controllers utilise the so-called parallel form of the PID equation (these are not the terms strictly used in academia and industry; sometimes equation (2.1) is referred to as ideal, for example). The parallel equation is given as follows:

$$W_c(s) = K_c \left(1 + \frac{1}{T_i s} + T_d s \right). \qquad (2.2)$$

In the parallel equation the tuning parameters are the proportional gain K_c, the integral time constant T_i and the derivative time constant T_d. There are a number of other forms of PID equations (see [75]). However, in this book only the above equations are used, and in most cases just the parallel equation (2.2). The obvious relationship between the two presented equations is as follows: the proportional gain is the same in both, the integral gain K_i is K_c/T_i and the derivative gain K_d is $K_c K_d$.

The task of tuning is to determine for a given process the values of K_c, T_i and T_d that provide optimal in a certain sense or acceptable performance of the control system (loop). The quality of tuning or loop performance can be determined based on the reactions of the closed loop to external control signals or disturbances, which may be intentionally created for the purpose of such a test or exist in the process (observations of the flow loop reactions to fluctuations of source pressure that naturally exist in the process, for example). There are a number of established and new methods of tuning, some of that were mentioned above. Below we consider non-parametric methods of tuning which fall into the category of methods based on the *continuous cycling*. The term "continuous cycling" is used to define the self-excited nonvanishing oscillations generated in the loop, which includes the process, aimed at measuring the parameters of these oscillations and producing the controller tuning parameters on the basis of the measurements obtained.

We will consider a few methods based on continuous cycling, presenting them in chronological order (at least the first three of them) and showing the development of the ideas as they are given sequentially in this overview.

2.2 Ziegler–Nichols Closed-Loop Test and Tuning

The open-loop and closed-loop tests proposed by J. Ziegler and N. Nichols [88] were the first methods in which a systematic and theoretically justified approach to controller tuning was introduced. Both methods—or at least some elements of both—are still used in practice and utilised in other methods of controller tuning. While the open-loop method is mainly related to parametric methods of loop tuning, the closed-loop method is purely a non-parametric method and related to other *continuous cycling* methods described in this book.

The methodology of closed-loop tuning proposed by Ziegler and Nichols is as follows. At the start the loop should be brought to a steady state. This can be done either manually through adjustments of valve position or by means of a PID controller through the use of the integral action. The controller might not be properly tuned but must deliver stability to the loop. It can have conservative gain values, which would provide a stable loop. At the second step, the integral and derivative components of the PID controller must be disabled and the loop should stay in the steady state. After that the proportional gain is incremented by steps and the behaviour of the loop is observed (see Figs. 2.1 and 2.2). Some sources recommend incrementing the gain by a value equal to half its current value. This, indeed, may speed up the process of tuning. Yet smaller increments must be applied once the tuning process approaches a sustained oscillation. What is observed is the occurrence of self-sustained oscillations. Because the loop is in a steady state, small increments of the set point up and down may be applied with attempts to excite an oscillation. If the oscillation vanishes (interval A of Fig. 2.2) then further increments of the proportional gain should be implemented. As shown for interval B, with an increase of the proportional gain, oscillations become less damped. And finally at some value of the gain they become divergent—as shown for interval C. At this point the gain value should be decreased to a safe level to exclude the possibility of significant process upset by an unstable loop. And, the gain value must be adjusted to obtain a nonvanishing self-sustained oscillation—as shown for interval D.

The Ziegler–Nichols test measures frequency and amplitude of oscillations, which are usually referred to as ultimate frequency Ω_u (or alternatively the ultimate period $T_u = \frac{2\pi}{\Omega_u}$) and ultimate gain K_u, which is the value of the proportional gain

Fig. 2.1 Ziegler–Nichols
closed-loop test

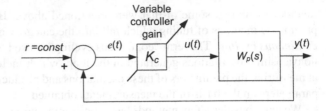

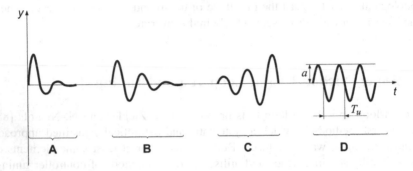

Fig. 2.2 Ziegler–Nichols closed-loop tuning. Steady self-excited oscillations are obtained in case D. Ultimate period $T_u = 2\pi/\Omega_u$ and ultimate gain are measured for this case

Table 2.1 Coefficients of tuning rules for Ziegler–Nichols closed-loop test

Controller	c_1	c_2	c_3
P	0.50		
PI	0.45	0.83	
PID	0.60	0.50	0.12

that provides nonvanishing self excited oscillations in the closed-loop system (option D in Figs. 2.2 and 2.3). In fact, oscillations in the loop are generated at the phase cross-over frequency ω_π of the process transfer function, which is the frequency at which the phase characteristic of the process is equal to $-\pi$ rad (or $-180°$).

The controller tuning parameters can be easily computed using the following formulas [88].

$$K_c = c_1 K_u, \qquad T_i = c_2 \frac{2\pi}{\Omega_u}, \qquad T_d = c_3 \frac{2\pi}{\Omega_u},$$

where c_1, c_2 and c_3 are coefficients defining the tuning rules. The coefficient values for the three types of controllers are presented in Table 2.1.

Tuning rules were developed to provide one quarter amplitude decay. This, however, is not always achieved in practice; Ziegler and Nichols determined the coefficient values using a specific process model. Moreover, even if this target requirement is met the result usually provides an aggressive tuning with an oscillatory response. However, the method provides a good starting point for subsequent fine tuning.

Fig. 2.3 Nyquist plots of Ziegler–Nichols closed-loop test

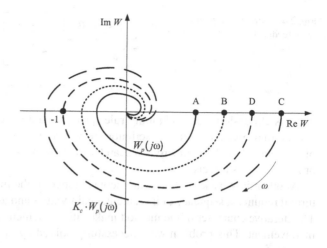

In terms of control theory, the closed-loop Ziegler–Nichols test is commonly interpreted by applying concepts linear systems analysis. This is a fairly simple and precise interpretation for most systems. There are some aspects of this test though that reveal effects which cannot be explained by linear theory. A respective example is considered in Chap. 4 of the book. The stepwise increase of the proportional gain can be seen as an expansion of the Nyquist plot of the process (Fig. 2.3). Each of the plots $K_c W_p(j\omega)$ in Fig. 2.3 can be associated with transients shown in Fig. 2.2. The correspondence is marked by the letters A, B, C and D. The Nyquist plots A and B do not encircle the point $(-1, j0)$ and correspond to vanishing oscillations in the system response to disturbance, while plot C encircles the point $(-1, j0)$ and corresponds to the diverging oscillations of the system output. And only Nyquist plot D goes through the point $(-1, j0)$ and, as a result, self-sustained nonvanishing oscillations exist in the system.

From Table 2.1, one can see that if, for example, a proportional controller is going to be used then the controller gain should be half the ultimate gain. And, the gain margin in this case is always guaranteed to be 2. This is a remarkable feature of the Ziegler–Nichols test. The desired gain margin of the proportional controller is guaranteed even without identification or any knowledge of the process parameters. This happens because measurement and tuning are realised in the same parametric space. Performance of the loop cannot be ensured, but it can be related to the gain margin, and the fact that stability is guaranteed is, of course, an advantageous and remarkable feature of the test. However, proportional control alone is seldom used in process control. PI and PID controllers are much more widely employed. And the noted property of guaranteed stability margins, unfortunately, cannot be applied to loops containing PI or PID controllers. This happens because of the shift of the ultimate frequency point of the Nyquist plot of the process from the real axis. This shift is due to the introduction of a PI or PID controller, which has a phase shift at frequency Ω_u. In the loop containing the PI/PID controller, the frequency ω_π does not coincide with the frequency Ω_u. Strictly speaking, we cannot say anything about the stability of such a system. Yet, in practice application of the Ziegler–

Fig. 2.4 Åström–Hägglund
relay feedback test

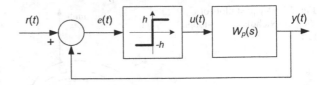

Nichols closed-loop test and tuning rules provides a good enough result. However, the test may give a little more oscillatory response than desired. This is a drawback of the method, and so numerous research efforts have been aimed at eliminating it or mitigating its effects.

Another drawback or at least an inconvenience of the method is its iterative nature: it requires a sequence of incremental gain values and associated test responses. The iterative character of the method makes its implementation in automatic tuners inconvenient. This problem was successfully solved by the tuning method considered next.

2.3 Åström–Hägglund Relay Feedback Test

In 1984 K. Åström T. and Hägglund proposed a test that was a substantial improvement of the Ziegler–Nichols test in terms of convenience of implementation in automatic loop tuners. They proposed replacement of the variable proportional gain with a nonlinear function and the use of the relay nonlinearity as this function (see Fig. 2.4). The patent application was first filed in Sweden in 1981 and later patented in a number of countries (in the United States, for example [40]). The patent specification included only the tuning rules of the Ziegler–Nichols closed-loop tuning method. However, later [5] other tuning rules were developed. The proposed test is now commonly known as the *relay feedback test* (RFT).

The idea of the test is based on the observation that the relay feedback system (Fig. 2.4) generates oscillations of the same frequency as the oscillations in the Ziegler–Nichols test. Yet, the test realisation is noniterative and particularly suitable for computer controlled systems (PLC or DCS). The equality of these two frequencies is, however, only approximate; the describing function (DF) method[1] normally applied in this analysis does not allow one to see the difference. But exact methods [5, 15, 80] allow one to find the difference. It is usually insignificant and in consideration of the Åström–Hägglund method we shall assume that these frequencies are equal.

In an analysis of the relay feedback system by the DF method, the nonlinear function can be replaced with an equivalent gain (complex or real), that is the describing function itself. This gain describes the propagation of the fundamental frequency component (first harmonic) in the Fourier series expansion of the periodic error

---

[1]It is also sometimes referred to as the sinusoidal input describing function; we shall omit sinusoidal in the subsequent text.

signal through the nonlinearity. For the DF method to be applied, the so-called *filtering hypothesis* (which states that the linear part of the system (process) must be a low-pass filter) must be valid. The describing function is not a constant value but a function of the amplitude and sometimes of the frequency of the input signal to the nonlinearity (error signal). For the ideal relay nonlinearity, the describing function $N(a)$ is given by the expression $N(a) = \frac{4h}{\pi a}$, where a is the amplitude of the oscillations of the error signal, and h is the amplitude of the relay. Once the relay is replaced with the DF the frequency of the self-excited oscillations can be found from the harmonic balance equation

$$N(a_0)W_p(j\Omega_0) = -1,$$

which can be interpreted as the Nyquist stability criterion applied to the system in which the relay is replaced with the DF. Due to inherent instability[2] the relay feedback system excites nonvanishing oscillations of the frequency Ω_0 and the amplitude a_0, which are measured in the test. Frequency Ω_0 is the ultimate frequency, and the ultimate gain is computed as the value of the DF: $K_u = N(a_0) = \frac{4h}{\pi a_0}$. Because the RFT was proposed as a further development of the Ziegler–Nichols method, initially its tuning rules were those of Ziegler and Nichols (see Table 2.1).

Later, Åström and Hägglund proposed other tuning rules, in [5], which is remarkable as the paper where an attempt to design tuning rules that would provide required specifications on gain or phase margin was made. However, this attempt was not fully successful because the tuning rules were designed in a way that would ensure the phase lag of the PID controller at the frequency of the test oscillations would be zero (the phase lead due to the derivative term is equalised by the same value of the phase lag introduced by the integral term):

$$\frac{1}{jT_i\Omega_0} + jT_d\Omega_0 = 0,$$

which leads to the constraint

$$T_iT_d = \frac{1}{\Omega_0^2}.$$

The drawback of this approach is obvious. The designed tuning rules could hardly provide optimal or near-optimal tuning; even the PI controller falls out of the class of controllers covered by this approach. Nevertheless, it was a remarkable vision of the real requirement and needs of control practice. Furthermore, the whole RFT was an ingenious design one that gave a boost to industry as it opened the way to the automation of controller tuning.

---

[2]From a theoretical point of view, zero can be a stable equilibrium point; for example it is an asymptotically stable equilibrium point in the relay system comprising an ideal relay and second-order linear dynamics [15].

2.4 Generating Test Oscillations in the Third Quadrant

Despite the innovative ideas given in [5], a popular notion has been that the most important point on system's frequency response is where the phase characteristic of the process equals to $-180°$ (frequency ω_π). This view appears in many publications. We shall refer to this point as the *phase cross-over frequency*. However, this is the most important point only in a system with a proportional controller, when introduction of the controller does not change the value of ω_π. This prerequisite is often neglected and the principle applied to all types of PID control. We consider the following motivating example and analyse how the introduction of the controller may affect the results of identification and tuning.

Example 2.1 Let us assume that a process is given by the following transfer function (which was used in a number of publications as a test process):

$$W_p(s) = \exp^{-2s} \frac{1}{(2s+1)^5}. \tag{2.3}$$

Find the first order plus dead time (FOPDT) approximating model $\hat{W}_p(s)$ to process (2.3) based on matching the values of the frequency response at frequency ω_π for the original and approximating models:

$$\hat{W}_p(s) = \frac{K_p e^{-s}}{T_p s + 1}, \tag{2.4}$$

where K_p is the process static gain, T_p is the time constant and τ is the dead time. Let us apply methods [88] to the tuning of process (2.3) and note that both (2.3) and (2.4) should produce the same ultimate gain and ultimate frequency in the Ziegler–Nichols closed-loop test [88] or the same values of the amplitude and the ultimate frequency in the RFT [5]. (Note: Strictly speaking, the values of the ultimate frequency in tests [88] and [5] are slightly different, as the frequency of the oscillations generated in the RFT does not correspond exactly to the phase characteristic of the process $-180°$; this fact follows from relay systems theory [15, 16, 80]; however, we shall use the describing function method and disregard inaccuracies.) Obviously, this problem has an infinite number of solutions, as (2.4) has three unknown parameters and only two measurements are obtained from the test. Assume the value of the process static gain is known: $K_p = 1$, and determine T_p and τ. These parameters can be found from equation

$$\hat{W}_p(j\omega_\pi) = W_p(j\omega_\pi),$$

where ω_π is the phase cross-over frequency for both transfer functions. Therefore, $\arg W_p(j\omega_\pi) = -\pi$. The value of ω_π is 0.283, which gives $W_p(j\omega_\pi) = (-0.498, j0)$, and the FOPDT approximation is, therefore, as follows (found via solution of the system of two algebraic equations):

$$\hat{W}_p(s) = \frac{e^{-7.393s}}{6.153s + 1}. \tag{2.5}$$

Fig. 2.5 Nyquist plots for process (2.3) and FOPDT approximation (2.5)

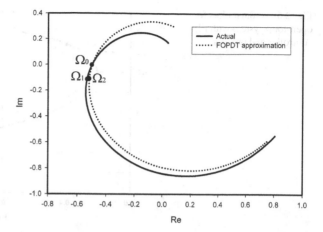

Fig. 2.5 Nyquist plots for process (2.3) and FOPDT approximation (2.5)

The Nyquist plots of process (2.3) and its approximation (2.5) are depicted in Fig. 2.5 (the meaning of frequencies Ω_1 and Ω_2 is explained below). The point of intersection of the two plots (denoted as Ω_0) is also the point of intersection with the real axis. Also $\Omega_0 = \omega_\pi$ for both process dynamics (2.3) and (2.5), and therefore $\hat{W}_p(j\Omega_0) = W_p(j\Omega_0)$. If the designed controller is of proportional type then the gain margins for processes (2.3) and (2.5) are the same. However, if the controller is of PI type then the stability margins for (2.3) and (2.4) are different. We illustrate this below. Design the PI controller given by the following transfer function:

$$W_c(s) = K_c\left(1 + \frac{1}{T_c s}\right),\qquad(2.6)$$

where K_c is the proportional gain and T_c is the integral time constant of the controller, using the Ziegler–Nichols tuning rules [88]. This results in the following transfer function of the controller:

$$W_c(s) = 0.803\left(1 + \frac{1}{17.76s}\right).\qquad(2.7)$$

The Nyquist plots of the open-loop systems containing process (2.3) or its approximation (2.5) and controller (2.7) are depicted in Fig. 2.6. It follows from the frequency-domain theory of linear systems and the tuning rules used that the mapping of point Ω_0 in Fig. 2.5 into point Ω_0 in Fig. 2.6 is done via clockwise rotation of vector $\hat{W}_p(j\Omega_0)$ by the angle $\psi = \arctan(1/(0.8 \cdot 2\pi)) = 11.25°$ and multiplication of its length by certain value, so that its length becomes equal to 0.408. This is possible due to the serial connection of the controller and the process and the possibility of treating their frequency response (at Ω_0) as vectors. However, for the open-loop system containing the PI controller, the points of intersection of the Nyquist plots of the system and of the real axis are different for the system with process (2.3) and with process approximation (2.5). They are shown as points Ω_1 and Ω_2 in Fig. 2.6. The points of frequencies Ω_1 and Ω_2 on the Nyquist plots of the original process and its approximation, respectively, are also shown in Fig. 2.5. Therefore, the stability margins of the systems containing a PI controller are no longer the same. They

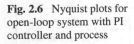

Fig. 2.6 Nyquist plots for
open-loop system with PI
controller and process

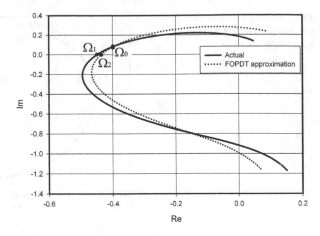

are revealed as different points of intersection of the plots and of the real axis in
Fig. 2.6. In fact, the position of vector $\hat{W}_{ol}(j\Omega_0) = \hat{W}_c(j\Omega_0)\hat{W}_p(j\Omega_0)$ is fixed, but
this vector does not reflect the stability of the system. As one can see in Fig. 2.6, the
gain margin of the system containing the FOPDT approximation of the process is
higher than the one of the system with the original process.

The considered example illustrates a fundamental problem of all methods of
identification-tuning based on the measurements of process response at the critical point (Ω_0). This problem is the shift of the critical point due to the introduction
of the controller. The question that follows from the above analysis is this: *Can the
test point be selected in a different way, so that the introduction of the controller
would be accounted for in the test itself*? And if this is possible, then *what kind of
test should it be to ensure the measurements in the desired test point*?

We address the first question now. Assume that we can design a certain test
so that we can generate the test frequency at the desired phase lag of the process
$\arg W_p(j\Omega_0) = \varphi$, where φ is a given quantity, and measure $W_p(j\Omega_0)$ in this point.
Consider the following example.

Example 2.2 Let the plant be the same as in Example 2.1. Assume that the introduction of the controller will give a mapping similar to the mapping described above—
the vector of the frequency response of the open-loop system at the point Ω_0 will be
a result of clockwise rotation of the vector $\hat{W}_p(j\Omega_0)$ by a known angle and multiplication by a certain known factor: $\hat{W}_{ol}(j\Omega_0) = \hat{W}_c(j\Omega_0)\hat{W}_p(j\Omega_0)$. Assume also
that the controller will be the same as in Example 2.1 (for illustrative purpose, since
the tuning rules are not formulated yet). Therefore, let us find the values of T_p and
τ for transfer function (2.4) (we still assume $K_p = 1$) that ensure that the equality
$\hat{W}_p(j\Omega_0) = W_p(j\Omega_0)$ holds, where $\arg W_p(j\Omega_0) = -180° + 11.25° = -168.75°$
(the angle is selected considering the subsequent clockwise rotation by $11.25°$).

Fig. 2.7 Use of delay for
identification in the third
quadrant

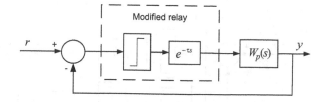

Therefore, $\Omega_0 = 0.263$ and $W_p(j\Omega_0) = (-0.532, -j0.103)$. The corresponding
FOPDT approximation of the process is

$$\hat{W}_p(s) = \frac{e^{-7.293s}}{5.897s + 1}. \qquad (2.8)$$

Application of controller (2.7) shifts the point Ω_0 of intersection of $W_p(j\Omega_0)$
and $\hat{W}_p(j\Omega_0)$ to the real axis. This point is still the point of intersection of the two
Nyquist plots. Therefore, the gain margin of both systems, with the original process
and with the approximated process, are the same. Consider now the problem of the
*design of a test that can provide matching the points of the actual and approximating
processes* at the point corresponding to a specified phase lag.

2.5 Tests that Ensure Frequency of Oscillations at Arbitrary Process Phase Lags

2.5.1 Test Using Additional Time Delay

There are a number of algorithms that can be used to generate oscillations in a
closed-loop test over the process at the frequency corresponding to the third quadrant of the process frequency response (Nyquist plot).

Tan et al. [78] suggested that an additional time delay should be included after
the relay nonlinearity to excite oscillations at the frequency lower than the phase
cross-over frequency (see also [79]). This delay can then be adjusted by iteration to
excite oscillations at the specified phase lag of the process. The proposed method
can be illustrated by the diagram as in Fig. 2.7.

The proposed algorithm involves finding two points on the Nyquist plot of the
process: at the ultimate frequency (phase cross-over frequency) and at the frequency
that delivers the specified phase lag of the process $-\pi + \phi_m$, where $\phi_m \in [0, \pi/2]$.

The algorithm includes the following iterations.

- Carry out the conventional RFT and find the ultimate gain K_u and the ultimate
  frequency Ω_u.
- With these values available, calculate the initial guess for the frequency $\hat{\omega}_\phi$ for the
  specified phase $-\pi + \phi_m$ as follows: $\hat{\omega}_\phi = ((\pi - \phi_m)/\pi)/\Omega_u$ and initialisation
  of the time delay $\tau = \phi_m/\hat{\omega}_\phi$.
- Continue the RFT with delay updating the delay value at each iteration as $\tau = \phi_m/\omega_{osc}$, where ω_{osc} is the frequency of the oscillations in the test.

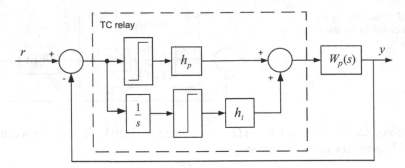

Fig. 2.8 Use of two-channel relay for identification in the third quadrant

The authors showed through simulations that the iterative algorithm converges to the frequency corresponding to the specified phase lag $-\pi + \phi_m$. The algorithm allows one to carry out identification for an arbitrary frequency point of the third quadrant of the complex plane but this is achieved at the expense of introducing iterations in the test, which significantly lengthen the test time and may cause unnecessary disturbance to the process.

2.5.2 Test Using Additional Integrator Term

Friman and Waller [38] proposed the use of the "two-channel relay" for identification of the process at a point of the third quadrant of the complex plane. It was proposed that an additional integrator having the error signal as the input and an additional relay, which output is added to the output of the main relay as depicted in Fig. 2.8, be used for exciting test oscillations.

An advantage of this test is that it does not require iteration, and the required phase lag can be obtained by the proper selection of the amplitudes of the two relays, as shown below.

The describing function of the *two-channel relay* can be computed as a sum of the two describing functions:

$$N(a) = \frac{4h_p}{\pi a} + \frac{1}{j\omega} \frac{4h_i}{\pi a_i},$$

where h_p and h_i are the amplitudes of the relays in the proportional link and the integral link, respectively, a is the amplitude of the oscillations (of the error signal), a_i is the amplitude of the oscillations after the integrator and ω is the frequency of the oscillations.

It is easy to see that $a_i = a/\omega$ and, therefore, the describing function of the algorithm can be rewritten as follows:

$$N(a) = \frac{4h_p}{\pi a} - j\frac{4h_i}{\pi a}.$$

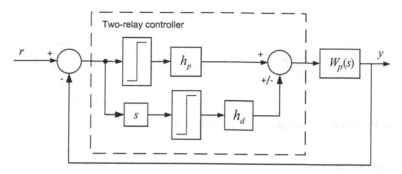

Fig. 2.9 Use of two-relay control for identification at the specified phase lag

Therefore, the magnitude of the describing function is

$$|N(a)| = \frac{4\sqrt{h_p^2 + h_i^2}}{\pi a}$$

and the phase response provided by the two-channel relay is

$$\arg N(a) = -\arctan \frac{h_i}{h_p}.$$

Overall the test is convenient and provides the required functionality but sometimes exhibits a long convergence time; in processes having nonlinearities (such as valve stiction) may result in low resolution in some points if the required $|h_p - h_i|$ difference is too small or too big.

2.5.3 Test Using Additional Derivative Term

A test that utilises the derivative of the error signal was proposed by Castellanos et al. [29]. The algorithm can be illustrated by Fig. 2.9. The algorithm originated from the analysis of the properties of the so-called *twisting* second-order sliding mode control algorithm. The original twisting control algorithm was modified to suit the purpose of identification in both ranges of the process phase lag: $[-\frac{\pi}{4}, -\frac{\pi}{2}]$ and $[-\frac{\pi}{2}, -\frac{3\pi}{2}]$:

$$N(a) = N_p(a) + j\omega N_d(a_d) = \frac{4h_p}{\pi a} + j\omega \frac{4h_d}{\pi a_d} = \frac{4}{\pi a}(h_p + jh_d), \qquad (2.9)$$

where h_p and h_d are the amplitudes of the relays in the proportional link and the derivative link, respectively, a is the amplitude of the oscillations (of the error signal), a_d is the amplitude of the oscillations after the differentiator and ω is the frequency of the oscillations.

The magnitude of the describing function is

$$|N(a)| = \frac{4\sqrt{h_p^2 + h_d^2}}{\pi a}$$

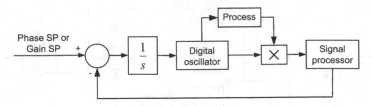

Fig. 2.10 Phase-lock loop non-parametric identification

and the phase response is

$$\arg N(a) = \arctan \frac{h_d}{h_p}.$$

Because $\arg N(a) > 0$ for $h_d > 0$, the two-relay algorithm provides identification in the range of process phase lags of $[-\frac{\pi}{2}, -\frac{3\pi}{2}]$. To change the range of the identification of the phase lags to $[-\frac{\pi}{4}, -\frac{\pi}{2}]$ the sign of h_d must be changed to negative.

The algorithm was initially designed to provide identification of electromechanical systems at frequencies higher than the phase cross-over frequency. Its use in process applications may feature high sensitivity to noise and, therefore, additional noise-reduction measures would be needed.

Exact analysis of oscillations in the system with the two-relay algorithm is presented in [18]; analysis of orbital stability of the periodic solution is given in [2] through the describing function method and in [1] through the construction of the linearised Poincaré map.

2.5.4 Test Using Phase-Lock Loop

A method aimed at the elimination of such known drawbacks of the relay feedback test as inaccuracy due to the use of the approximate describing function method and identification in only one point corresponding to the phase cross-over frequency was proposed by Crowe and Johnson [33]. The method is based on the phase-lock principle involving the organisation of a loop that includes a process. Oscillations are excited in this loop at the frequency that would ensure the desired phase lag. The block diagram of the system is given in Fig. 2.10.

The proposed identifier includes the following components (as per [34]):

- A feedback loop using a phase or gain reference.
- A digital controlled oscillator providing a sinusoidal signal.
- A multiplier.
- A digital signal processing unit, which computes the actual phase lag or gain of the process.
- A digital integrator, which is supposed to eliminate nonzero error and provide convergence to the steady state in the loop.

While providing such advantages as high accuracy of identification and the possibility of identification at a few frequency points, which motivated this method's development, this approach also features some drawbacks such as higher complexity, longer convergence time required for the phase-lock loop to come to a steady state and disconnection of the process from the control loop for the duration of the test. The results of the comparison of the phase-lock loop identification against the conventional RFT method, as well as details of implementation and accuracy analysis, are presented in [34], [35] and [36].

2.6 Conclusions

A brief overview of non-parametric methods was presented in this chapter, starting with the Ziegler–Nichols closed-loop test. Evolution of the ideas used in non-parametric tuning is shown through the review of methods. It is shown that the use of test oscillations of the frequency that corresponds to the $-180°$ phase response of the process does not allow one to design PI or PID controller tuning loops that would ensure specified gain or phase margins. This happens due to the shift of the test point from the real axis when the controller is introduced in the loop. It is also shown that generation of test oscillations in the third quadrant of the complex plane would be beneficial. A number of tests providing test oscillations in the third quadrant are reviewed. Another method of PID controller tuning, in which coordinated test and tuning are combined, is presented in Chap. 3.

Chapter 3
Modified Relay Feedback Test (MRFT) and Tuning of PID Controllers

The holistic approach to PID controller tuning, which involves the coordinated use of modified relay feedback test and tuning rules, is presented in this chapter. It is shown that with this approach, specifications to gain margin or phase margin can be satisfied. Performance requirements to a number of typical industrial processes are analysed, and optimal tuning rules for the modified relay feedback test application are produced on the basis of this analysis.

3.1 MRFT and Holistic Approach to Test and Tuning

3.1.1 Modified Relay Feedback Test

A number of tests that ensure generation of test oscillations at the specified process phase lag were considered in Chap. 2. Another test, a modification of the RFT, is presented below.

For a given process, consider the following discontinuous control, which is also referred to as the *modified relay feedback test* (MRFT):

$$u(t) = \begin{cases} h & \text{if } e(t) \geq b_1 \text{ or } (e(t) > -b_2 \text{ and } u(t-) = h) \\ -h & \text{if } e(t) \leq -b_2 \text{ or } (e(t) < b_1 \text{ and } u(t-) = -h) \end{cases} \tag{3.1}$$

where $b_1 = -\beta e_{min}$, $b_2 = \beta e_{max}$, $e_{max} > 0$, $e_{min} < 0$ are the last "singular" points of the error signal (Fig. 3.1) corresponding to the last maximum and minimum values of $e(t)$ after crossing the zero level; $u(t-) = \lim_{\epsilon \to 0, \epsilon > 0} u(t - \epsilon)$ is the control value at the time immediately preceding current time t; h is the amplitude of the relay; β is a constant. Initial values of e_{max} or e_{min} can be assigned as $e(0)$ (the choice of either e_{max} or e_{min} depends on the sign of $e(0)$). Formula (3.1) provides a discontinuous control algorithm (the switching is defined by the values of b_1 and b_2, which in turn depend on the singular points e_{min} and e_{max}) that is similar to the so-called "generalised suboptimal" algorithm used for generating a second-order sliding mode in systems of relative degree two [10, 11]. The difference between

I. Boiko, *Non-parametric Tuning of PID Controllers*, Advances in Industrial Control, DOI 10.1007/978-1-4471-4465-6_3, © Springer-Verlag London 2013

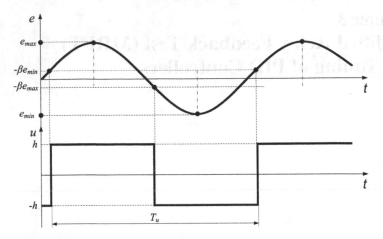

Fig. 3.1 Modified relay feedback test signals

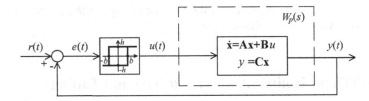

Fig. 3.2 Modified relay feedback test

the algorithms is this: The generalised suboptimal algorithm involves an advance switching of the relay (in the algorithm formulation (3.1), parameter $\beta < 0$), which is aimed at driving the system states to zero, whereas the proposed algorithm is aimed at generating a limit cycle and normally involves a lagged switching of the relay (parameter $\beta > 0$; however, the advance switching is possible, too), in which the lag value is coordinated with controller tuning rules.

Let the reference signal $r(t)$ in the system of Fig. 3.2 be zero. We show now that in the steady mode the motions in the system, where the control is given by (3.1) are periodic. Apply the describing function (DF) method [8] to the analysis of motions in the system Fig. 3.2. DF is an approximate method but—as confirmed by the practice of applications of RFT in process control—provides a model of satisfactory accuracy. Assume that the steady mode is periodic and prove this is a valid assumption by finding parameters of this periodic motion. If the motions in the system are periodic then e_{max} and e_{min} represent the amplitude of the oscillations: $a_0 = e_{max} = -e_{min}$ and the equivalent hysteresis value of the relay is

Fig. 3.3 Finding periodic solution

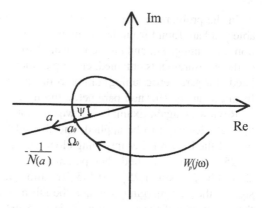

$b = b_1 = b_2 = \beta e_{max} = -\beta e_{min}$. The DF of the hysteretic relay is given as the following function of amplitude a [8]:

$$N(a) = \frac{4h}{\pi a}\sqrt{1 - \left(\frac{b}{a}\right)^2} - j\frac{4hb}{\pi a^2}, \quad a > b. \tag{3.2}$$

However, system Fig. 3.2 with control (3.1) is not a conventional relay system. It has an hysteresis value that is unknown a priori and depends on the amplitude value: $b = \beta a$. Therefore, (3.2) can be rewritten as follows:

$$N(a) = \frac{4h}{\pi a}\left(\sqrt{1 - \beta^2} - j\beta\right). \tag{3.3}$$

The MRFT will generate oscillations in the system under control (3.1). Parameters of the oscillations can be found from the harmonic balance equation:

$$W_p(j\Omega_0) = -\frac{1}{N(a_0)}, \tag{3.4}$$

where a_0 and Ω_0 are the amplitude and the frequency of the periodic motions. The negative reciprocal of the DF is given as follows:

$$-\frac{1}{N(a)} = -\frac{\pi a}{4h}\left(\sqrt{1 - \beta^2} + j\beta\right). \tag{3.5}$$

Figure 3.3 gives a simple graphic interpretation for finding a periodic solution of the system in Fig. 3.2 with control (3.1). The graph shows the Nyquist plot of the process and the negative reciprocal of the describing function. The DF is the straight half-line in the third quadrant, lying at angle $\psi = \arcsin \beta$ below the real axis. The periodic solution sought is at the point of intersection of these two plots. The solution exists only if the Nyquist plot of the process lies in the third quadrant of the complex plane. Due to inevitable existence of small delays in process dynamics, this condition is always satisfied in practice. But in addition to this, the periodic solution must satisfy a condition of orbital stability. Orbital stability may take place in a system having stable or unstable processes.

In the problem of analysis, frequency Ω_0 and amplitude a_0 are unknown variables and are found from the complex equation (3.4). In the problems of identification and tuning, Ω_0 and a_0 are measured from the modified RFT, and on the basis of the measurements obtained, either parameters of the *underlying* model are calculated (for parametric tuning) or controller tuning parameters are calculated directly from Ω_0 and a_0 (for non-parametric tuning).

Reviewing again Example 2.2, we can note that if, for example, Ziegler–Nichols tuning rules were to be applied, and the subsequent transformation via introduction of the PI controller involving clockwise rotation by angle $\psi = \arctan \frac{1}{0.8 \cdot 2\pi} = 11.25°$ was to be applied, then parameter β of the controller for the modified RFT should be $\beta = \sin 11.25° = 0.195$. The modified RFT also allows for the exact design of the gain margin (assuming the DF method provides an exact model). Since the amplitude of the oscillation a_0 is measured from the test, the process gain at frequency Ω_0 can be obtained as follows:

$$\left| W_p(j\Omega_0) \right| = \frac{\pi a_0}{4h}, \tag{3.6}$$

which after introduction of the controller will become the process gain at the critical frequency.

We should note, however, that, although it is very simple, control (3.1) is an algorithm. This algorithm can be detailed as the following steps:

- Step 1. Assign $e_{min} = e_{max} = 0$.
- Step 2. Compute $b_1 = -\beta e_{min}$ and $b_2 = -\beta e_{max}$ (which both are *zero* in the initial point).
- Step 3. Compute the control in accordance with (3.1) and apply the control signal to the process.
- Step 4. Measure the process variable and compute the error signal $e(t)$. Once the error signal reaches minimum or maximum, reset the value of either e_{min} or e_{max}, respectively, assigning it to the minimum or maximum of $e(t)$, and go to Step 2.
- Step 5. Stop the test once periodic motion has been established (i.e. amplitude and frequency of the oscillations are stabilised) and measurements of frequency and amplitude of the test oscillations have been made; or a preset maximum number of cycles has been generated; or the maximum time allocated for the test has elapsed.

3.1.2 Homogeneous Tuning Rules

It is shown above that test oscillations can be generated at a point in the third quadrant of the complex plane using the modified RFT. However, to use this advantage to ensure the desired gain or phase margin, the tuning rules must be coordinated with the test as follows.

Let the PID controller transfer function be

$$W_c(s) = K_c \left(1 + \frac{1}{T_i s} + T_d s \right).$$

Define the following tuning rules format of a PID controller:

$$K_c = c_1 \frac{4h}{\pi a_0}, \qquad T_i = c_2 \frac{2\pi}{\Omega_0}, \qquad T_d = c_3 \frac{2\pi}{\Omega_0}, \qquad (3.7)$$

where c_1, c_2 and c_3 are constant parameters that define the tuning rule, to which we shall refer as to the *homogeneous tuning rules*, because the tuning parameters are homogeneous functions of critical gain (given by $\frac{4h}{\pi a_0}$) and critical period (inverse critical frequency). The idea behind tuning rules (3.7) is that we may scale the tuning parameters for processes having different time constants. We note that if the tuning rules are given by (3.7) then the closed-loop system characteristics become invariant to the time constants of the process, so that if all time constants of the process were increased by the factor ρ then the critical frequency would decrease by the same factor ρ, and the product of every time constant by the critical frequency would remain unchanged. The use of the *homogeneous tuning rules* along with the modified RFT will allow us to fully utilise the features of non-parametric tuning methods. If *homogeneous tuning rules* (3.7) are used then the frequency response of the PID controller at frequency Ω_0 becomes

$$W_c(j\Omega_0) = c_1 \frac{4h}{\pi a_0} \left(1 - j \frac{1}{2\pi c_2} + j 2\pi c_3 \right) \qquad (3.8)$$

Therefore, if the tuning rules are established through the choice of parameters c_1, c_2 and c_3, and the test provides self-excited oscillations of frequency Ω_0, which will be equal to the phase cross-over frequency ω_π of the open-loop system (including the controller), then the controller phase lag at the frequency $\omega_\pi = \Omega_0$ will depend only on the values of c_2 and c_3:

$$\varphi_c(\omega_\pi) = \arctan \left(2\pi c_3 - \frac{1}{2\pi c_2} \right), \qquad (3.9)$$

which directly follows from formula (3.8) if $\omega_\pi = \Omega_0$.

3.1.3 Non-parametric Tuning Rules for Specification on Gain Margin

It seems that the most appealing application of the presented test is a non-parametric tuning. However, given a large variety of possible process dynamics, it is difficult to formulate certain universal rules for tuning. In the practice of process control, tuning rules that provide a less aggressive response than the one provided by the various integral performance criteria or Ziegler–Nichols formulas (and others) are widely used. This approach is motivated by a consideration of safety, which is chosen over high performance. This trend is reflected in the review of modern PID control given in [7].

We derive now the relationship that would allow us to tune PID controllers with specification on gain margin for the open-loop system. Let the specified gain margin be $\gamma_m > 1$ (in absolute value). Then, taking the absolute value of both sides of (3.8) and considering (3.6), we obtain the following equation:

$$\gamma_m c_1 \sqrt{1 + \left(2\pi c_3 - \frac{1}{2\pi c_2}\right)^2} = 1, \tag{3.10}$$

which is a constraint that is complementary to tuning rules (3.7). To provide the specified gain margin, the modified RFT must be carried out with parameter

$$\beta = - \sin \varphi_c(\Omega_0)$$

$$= - \sin \arctan\left(2\pi c_3 - \frac{1}{2\pi c_2}\right)$$

$$= - \frac{2\pi c_3 - \frac{1}{2\pi c_2}}{\sqrt{1 - (2\pi c_3 - \frac{1}{2\pi c_2})^2}}. \tag{3.11}$$

In the example above, if we keep parameter c_2 the same as [88] ($c_2 = 0.8$), then to obtain, for example, gain margin $\gamma_m = 2$ the tuning parameter c_1 for the modified RFT should be selected as $c_1 = 0.49$, and parameter β for the test should be selected in accordance with (3.11) as $\beta = 0.195$. For any process, the system will have gain margin $\gamma_m = 2$ (6 dB) exactly (within the framework of the filtering hypothesis of the DF method). Therefore, the modified RFT with parameter β calculated as (3.11) and tuning rules (3.7) satisfying constraint (3.10) can ensure the desired gain margin. However, (3.10) is an equation containing three unknown variables, which gives one freedom to vary parameters c_1, c_2 and c_3. We do not consider in this section the problem of optimal selection of these parameters. However, simple tuning rules can be easily obtained with the use of some characteristics of the tuning rules of [88]. In particular, we assume that the controller should provide the same phase response on the frequency of oscillation of the modified RFT $\varphi_c(\Omega_0)$ as the phase response of the corresponding controller at the critical frequency of conventional RFT tuned in accordance with the rules [88]. As a result, we can use values of parameters c_2, c_3 equal to the corresponding values of Table 2.1. Coefficients defining the tuning rules for gain margin $\gamma_m = 2$ in the format of values of parameters c_1, c_2, c_3 along with the values of $\varphi_c(\Omega_0)$ and parameter β for the test are given in Table 3.1.

One should note the difference between the values of the critical frequency of the conventional RFT and the frequency of oscillations in the modified RFT (except for the proportional controller). Therefore, even if the coefficients c_2, c_3 of Table 3.1 have the same values as corresponding coefficients of [88], they will actually produce different values of controller parameters T_i and T_d. In fact, due to the negative value of $\varphi_c(\Omega_0)$ for the PI controller (and consequently, lower frequency of oscillations of the modified RFT), one would get a higher value of T_i computed through the modified RFT and data of Table 3.1. And vice versa, due to the positive value of $\varphi_c(\Omega_0)$ for the PID controller, one would get lower values of T_i and T_d computed through the modified RFT and data of Table 3.1.

Fig. 3.4 Modified RFT and
tuning with specification on
phase margin

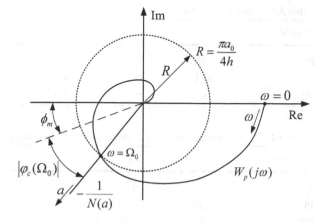

3.1.4 Non-parametric Tuning Rules for Specification on Phase Margin

We derive now the relationship that would allow us to tune PID controllers with specification on phase margin for the open-loop system. Using the same format of homogeneous tuning rules (3.7), and considering that if the parameter β of the modified RFT is calculated from the sum of $\varphi_c(\Omega_0)$ and the phase margin ϕ_m as:

$$\beta = \sin\big(\phi_m - \varphi_c(\Omega_0)\big) = \sin\left(\phi_m + \arctan\left(\frac{1}{2\pi c_2} - 2\pi c_3\right)\right), \qquad (3.12)$$

we formulate the constraint for the tuning rules ensuring ϕ_m as follows:

$$c_1\sqrt{1 + \left(2\pi c_3 - \frac{1}{2\pi c_2}\right)^2} = 1. \qquad (3.13)$$

The graphical interpretation of the modified RFT and tuning with specification on phase margin are presented in Fig. 3.4.

Indeed, if tuning rules (3.7) are subject to constraint (3.13) then at frequency Ω_0 of the modified RFT: (a) the absolute value of the open-loop frequency response, in accordance with (3.6) and (3.8), is

$$\big|W_{ol}(j\Omega_0)\big| = \big|W_c(j\Omega_0)\big|\big|W_p(j\Omega_0)\big|$$

$$= \frac{\pi a}{4h}c_1\frac{4h}{\pi a}\sqrt{1 + \left(2\pi c_3 - \frac{1}{2\pi c_2}\right)^2}$$

$$= c_1\sqrt{1 + \left(2\pi c_3 - \frac{1}{2\pi c_2}\right)^2} = 1,$$

which constitutes the *magnitude cross-over frequency*, and (b) the phase of the open-loop frequency response is $\arg W_{ol}(j\Omega_0) = \arg W_c(j\Omega_0) + \arg W_p(j\Omega_0) =$

Table 3.1 Coefficients defining sample tuning rules for gain margin $\gamma_m = 2$

Controller	c_1	c_2	c_3	$\varphi_c(\Omega_0)$	β
P	0.50	0	0	0	0
PI	0.49	0.80	0	$-11.2°$	0.195
PID	0.46	0.50	0.12	$23.5°$	-0.399

Table 3.2 Coefficients defining sample tuning rules for phase margin $\phi_m = 45°$

Controller	c_1	c_2	c_3	$\varphi_c(\Omega_0)$	β
P	1.0	0	0	$0°$	0
PI	0.98	0.80	0	$-11.2°$	0.831
PID	0.92	0.50	0.12	$23.5°$	0.367

$-180° + (\phi_m - \varphi_c(\Omega_0)) + \varphi_c(\Omega_0) = -180° + \phi_m$, which shows that the specification on the phase margin is satisfied. Again, assuming that the controller at frequency Ω_0 of the modified RFT should provide the same phase response as occurs at the critical frequency of RFT (which, of course, does not give optimal tuning rules, and those tuning rules are provided only to illustrate a possible option), we can obtain the values of parameters c_1, c_2 and c_3 seen in Table 3.2 for $\phi_m = 45°$. As in tuning with specification on gain margin, one should note in phase margin tuning the difference between critical frequency of the conventional RFT and the frequency of oscillations in the modified RFT, which yields different values of the controller parameters.

It is noted above that the derived relationships (3.10), (3.13) among the coefficients c_1, c_2, c_3 are not tuning rules yet. The problem of development of optimal tuning rules satisfying constraints (3.10) or (3.13) remains unsolved; different plant/process dynamics require different optimal tuning rules. Tables 3.1 and 3.2 provide examples of not optimal but satisfactory tuning rules that were generated on the basis of [88] by keeping the same values of c_2 and c_3 and finding c_1 that would satisfy (3.10) or (3.13). Optimal tuning rules may be produced for a particular class of plant/process dynamics via solving the problem of parametric optimisation for c_1, c_2, c_3 with a certain criterion (an integral performance criterion or other) and constrains (3.10) and (3.13). The solution of this problem is presented below.

The controller tuning can be described as the following step-by-step algorithm:

- Step A. The type of controller (P, PI or PID) and the tuning rules (for example, given by either Table 3.1 or Table 3.2, or generated in a similar way for other values of gain/phase margin) are selected.
- Step B. The modified RFT with parameter β corresponding to the selection made in step A is carried out.
- Step C. The values of frequency Ω_0 and amplitude a_0 of the self-excited oscillations in the system are measured.
- Step D. Tuning parameters of the controller are calculated per (3.7).

Fig. 3.5 Nyquist plots of
open-loop systems for
Example 3.1

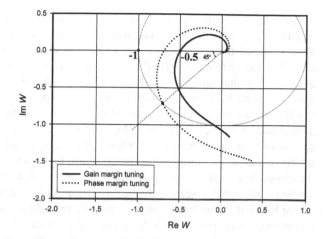

3.1.5 Example

Example 3.1 Consider the process transfer function (2.3) used in Example 2.1.
(a) Apply the modified RFT with amplitude $h = 1$, parameter $\beta = 0.195$ and c_1,
c_2 values from Table 3.1 for tuning a PI controller with specification on gain margin
$\gamma_m = 2$. (b) Use the modified RFT with amplitude $h = 1$, parameter $\beta = 0.831$ and
c_1, c_2 values from Table 3.2 for tuning a PI controller with specification on phase
margin $\phi_m = 45°$. The controller tuning that is done according to the presented
method produces the following results: (a) The modified RFT gives $\Omega_0 = 0.263$
and $a_0 = 0.691$; for tuning with specification on gain margin the controller parameters calculated per (3.7) are $K_c = 0.903$, $T_{ic} = 19.11$; (b) The modified RFT gives
$\Omega_0 = 0.187$ and $a_0 = 0.918$; for tuning with specification on gain margin the controller parameters calculated per (3.7) are $K_c = 1.359$, $T_{ic} = 26.88$. The frequency
response of the open-loop systems is presented in Fig. 3.5. One can see that, indeed,
the gain margin is 2 for option "a" (solid line), and the phase margin is 45° for
option "b" (dotted line).

3.2 Process-Specific Optimal Tuning Rules

3.2.1 General Approach to Producing Process-Specific Optimal Tuning Rules

In the previous section on the modified RFT, we designed a test and established tuning rules that allow one to ensure the specified gain margin or phase margin of the
system. However, the tuning rules were established only in terms of the equations
relating the tuning coefficients c_1, c_2 and c_3. Therefore, there is some freedom in assigning coefficients c_1 and c_2 for a PI controller or c_1, c_2 and c_3 for a PID controller.

As long as the combination of coefficients determining the tuning rules satisfies equation (3.10) for the gain margin tuning or equation (3.13) for the phase margin tuning and the test is carried out with parameter β satisfying conditions (3.11) or (3.12), the system will have the specified gain margin or phase margin. Yet, it is quite obvious that some combinations of the coefficients satisfying (3.10) or (3.13) would provide better results in terms of step response, for example, than others. For that reason the problem of finding optimal, in a certain sense, tuning rules, or a combination of the coefficients which define the tuning rule, is quite meaningful and its solution would benefit the practice of process control. The characteristic that might be used as an optimisation criterion can, for example, be the step response of the closed-loop system. However, it is also obvious that the solution of the optimisation problem makes sense only for certain particular types of process or plant. For example, what is optimal for flow loops may provide a poor result for level loops, and so on. Yet despite the fact that all flow processes differ from one another, they differ even more from level processes. This is because all flow processes are self-regulating and have similar values of the time constants, while all level processes are integrating and usually have quite a different order of the time constants compared to the typical flow process.

To mathematically define the optimisation problem we need to establish the criterion of optimisation (cost function), determine the constraints and select the optimisation method. The constraints naturally come from the MRFT method as equations (3.10) and (3.13) providing the equality constraint from the specification on either gain margin or phase margin, with respective determination of MRFT parameter β. The criterion of optimisation must be chosen as a function from a domain other than the frequency domain because otherwise the constraint as a frequency-domain function would have a high correlation with the optimisation criterion. For example, although we could theoretically maximise the phase margin subject to the fulfilment of the specified gain margin, this would hardly be a good formulation of the optimisation problem: Usually, imposing constrains on one frequency-domain characteristic leaves very little room for varying the other characteristic. It would be more expedient and productive to select the optimisation criterion from the time domain.

Time-domain criteria of optimality are very popular in the practice of process control and other industrial controls. Their popularity is related to the fact that the controller performance can be easily assessed using process trends through such performance metrics as response time, overshoot, etc. The most popular time-domain criteria used in optimisation are the so-called *integral performance criteria* measured through the process reaction to the step input signals. The following criteria are often used:
integral absolute error (IAE)

$$Q_{IAE} = \int_0^\infty |e(t)|\, dt, \tag{3.14}$$

integral square error (ISE)

$$Q_{ISE} = \int_0^\infty e^2(t)\, dt, \tag{3.15}$$

integral time absolute error (ITAE)

$$Q_{ITAE} = \int_0^\infty t|e(t)|\,dt, \qquad (3.16)$$

integral time square error (ITSE)

$$Q_{ITSE} = \int_0^\infty te^2(t)\,dt, \qquad (3.17)$$

and some others.

Non-integral time domain cost functions are used too. For example, one such criterion is the minimum settling time. Sometimes, in addition to the cost function, an inequality-type constraint is added, such as *minimum settling time* subject to the *constraint on the overshoot* value [17]:

$$Q_{set} = t_{set}(K_c, T_i, T_d),$$
$$\max_{t \in [0,\infty]} y(t) \le 1 + y_{os}/100,$$

where y_{os} is the maximum overshoot specification [%],

$$t_{set} = \begin{cases} y(t_{set}) = 1 - \Delta \quad \text{or} \quad y(t_{set}) = 1 + \Delta \\ y(t > t_{set}) \in [1 - \Delta, 1 + \Delta] \end{cases}$$

is the settling time, and 2Δ is the step response envelope width.

It is also worth noting that the step signal can be applied either as the set point or as a disturbance. And strictly speaking, optimal tuning rules generated from the consideration of these two different step responses would be different.

Another aspect of the considered optimisation problem is that the optimisation has to be done not for one particular process but for a set of distinct processes of the same kind—representing possible variations of gains and time constants within a certain model structure. We shall treat gains and time constants of a process model as certain *situational parameters* that define possible process dynamics, whereas tuning coefficients c_1, c_2 and c_3 are optimisation parameters. This situation gives a few approaches to optimisation.

Case A. The simplest is to produce a certain typical or averaged model of the process and carry out optimisation for the parameters of the tuning rules with respect to this model. It is important though that this "averaging" be done correctly because different parameters of the model may often be highly correlated. This correlation does not allow one to find the "averaged" model merely through averaging the parameters over the ranges of their variation.

Case B. The consideration of the worst case for producing the cost function value used in the optimisation is another approach. However, even if the process model variations are given through the ranges of model parameters, finding the worst case may become a complex problem because dependence of the cost function on a model parameter is not necessarily seen as a monotonic function. If the model has a few parameters, which is normally the case, the search for the "worst case" may become computationally cumbersome because the comparison of each point giving a combination of the model parameters with other points is done on the basis of solving of the optimisation problem for this point. The solution can be simplified if the number of model parameters is reduced through normalisation (this approach is illustrated below).

Case C. A probabilistic approach is valid too. It allows one to utilise information about the probability of the occurrence of the particular situation representing a combination of model parameters (*situational*), so that the priority is given to situations more likely to happen. We should emphasise here that the conditions of either required gain or phase margins remain intact, and the produced tuning rules will provide good tuning for all combinations of the model parameters. However, only those which occur most frequently will have tuning close to optimal.

Case D. An "equalizing" approach was proposed in [22] as a criterion of optimisation. According to this approach, what is minimised is not the value of the integral (or other) cost function but the difference between the maximum and the minimum values of the cost function obtained over the range of model parameters. This approach would not be feasible without the equality constraints giving the required gain or phase margins, because the optimisation might produce some meaningless result—for example, extremely low proportional gain—or it might just make the optimisation process diverging. Having constraints prevents this from happening. The whole criterion is aimed, therefore, at making tuning possibly uniform on the specified set of model parameters. The characteristics of tuning (aggressive, moderate, sluggish) can be controlled through the choice of the gain or phase margin.

3.2.1.1 Selection of Implied Process Model for Optimisation

We shall call the process model that is used for finding optimal tuning rules through the solution of the parametric optimisation problem (nonlinear programming problem) the *implied process model*. The implied process model is different from the underlying process model that is used in methods of identification and parametric tuning. The underlying process model is usually chosen from consideration of sufficient simplicity, subject to retaining some important properties of the process. The requirement of simplicity comes from limitations of the number of tests on the process, their accuracy and available computing power. There is also a problem of the nontrivial dependence of the final result of controller tuning and system performance on the model complexity, which is not a monotonic function. Therefore, a good rule of thumb is to select the underlying process model simple enough while retaining the main features of the actual process behaviour.

As mentioned, the implied process model is not used for identification of parameters, and it can be quite complex. The only purpose of this model is one of finding optimal tuning rules. In fact, the more precise and adequate this model is the more accurate the tuning rules that can be obtained. Therefore, there are good reasons to use more accurate models of the processes. Then, when the derived tuning rules are applied to the actual process, the tuning will be more accurate.

In our selection of the implied process model we are not limited by the order and, quite the contrary, we are motivated to select as precise a model as possible; thus, for basic processes like the fluid flow process instead of the commonly used

Fig. 3.6 Process output in RFT for FOPDT and DSOPDT models

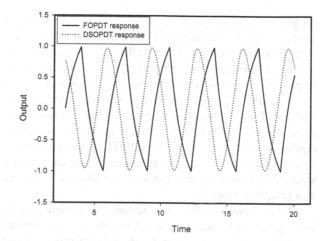

first-order plus dead time (FOPDT) model $W(s) = \frac{e^{-\tau s}}{Ts+1}$, we are going to use the second-order plus dead time (SOPDT) model:

$$W_p(s) = \frac{K_p e^{-\tau s}}{T^2 s^2 + 2\xi T s + 1}. \qquad (3.18)$$

This order increase serves the purpose of bringing the process response in the MRFT in compliance with the used method of analysis, which is the describing function method that is used under the assumption of the filtering hypothesis. According to this hypothesis, the output of the process in the modified RFT must be sinusoidal (in a steady mode). Obviously, it is impossible to achieve this with the FOPDT model, while easy to have when using the SOPDT model (see Fig. 3.6). However, the SOPDT model has not three but four parameters: static gain, time constant, dead time (delay) and damping coefficient, which creates the situation of an additional degree of freedom. We will approach this situation considering the purpose of the use of the higher-order model, which is bringing the response of the plant in correspondence with the limitations of the analysis method, and the fact that most of processes in the process control industry are overdamped.

We select the value of the damping coefficient corresponding to a damped response of the SOPDT process model. The exact selection of the damping coefficient is not very important—as long as the model provides a damped and smooth (close to sinusoidal) response in the MRFT—because the model is used in the nonparametric tuning and must reproduce just certain qualitative properties of an actual process. Strictly speaking, the value of the damping coefficient we are going to use is smaller than 1, which provides an underdamped response. The choice in favour of smaller values of the damping coefficient is motivated by the better filtering properties of this model. However, the damping with this value is still high enough, so we shall refer to this as a *damped* response. Because of the high damping properties of the SOPDT model we shall refer to it as the *damped second-order plus dead time* (DSOPDT) model. We shall also establish some correspondence between the DSOPDT model and the FOPDT model in terms of providing the same values of the phase characteristic at the phase cross-over frequency (for RFT) or the same frequency of the oscillations at MRFT (subject to the filtering hypothesis). This would help us quantify the ratio between the delay present in the process dynamics and the dominant time constant of these dynamics. It would also correspond to the practice of process control in which the step test remains the most widely used test over the process, allowing one to determine the dead time and the time constant from the *S-curve* of the step response. And, even if the detailed process model is more complex than the FOPDT model, knowledge of the time constant, which would be an equivalent time constant of the FOPDT model, is helpful and convenient since the dynamics of many processes are quantified in terms of the dead time-to-time constant ratio. Following this approach, we also find some equivalence of the FOPDT and DSOPDT models.

The conventional RFT oscillations are generated at the phase cross-over frequency, i.e. the frequency at which the phase characteristic of the open-loop system is equal to $-180°$. Therefore, for the FOPDT process given by the transfer function

$$W_p(s) = \frac{K_p e^{-\tau s}}{T_e s + 1},\qquad(3.19)$$

where T_e is the time constant of the FOPDT model equivalent to the DSOPDT model, the phase cross-over frequency ω_π is found from the following equation:

$$-\tau\omega_\pi - \arctan(T_e\omega_\pi) = -\pi.\qquad(3.20)$$

And for the DSOPDT model given by the transfer function $W(s) = \frac{e^{-\tau s}}{T^2 s^2 + 2\xi T s + 1}$, the equation that determines the phase cross-over frequency can be written as

$$-\tau\omega_\pi - \arctan\left(\frac{2\xi T\omega_\pi}{1 - T^2\omega_\pi^2}\right) = -\pi.\qquad(3.21)$$

Equating left-hand sides of (3.20) and (3.21) yields

$$T_e\omega_\pi = \frac{2\xi T\omega_\pi}{1 - T^2\omega_\pi^2}.\qquad(3.22)$$

Now, with this equivalence established, computation of T and T_e can be done as follows. If what is known is the FOPDT model and we must find the equivalent DSOPDT model, equation (3.20) must be solved for ω_π first; then the time constant of the DSOPDT model is found through the solution of equation (3.22), which is a quadratic equation for T, as follows:

$$T = \frac{-\xi + \sqrt{\xi^2 + T_e^2\omega_\pi^2}}{T_e\omega_\pi^2}.\qquad(3.23)$$

If, conversely, the DSOPDT model is known and we have to find the parameters of the FOPDT model, then for the given τ and T, equation (3.21) must be solved for ω_π first, after which the time constant T_e is found from (3.22) as follows:

$$T_e = \frac{2\xi T}{1 - T^2\omega_\pi^2}.\qquad(3.24)$$

Therefore, the relationship between T of the DSOPDT model and T_e of the FOPDT model for the conventional RFT can be found a priori, which is presented in Table 3.3.

Table 3.3 Equivalence between FOPDT and DSOPDT models for RFT

τ/T	1.72	1.86	2.00	2.14	2.28	2.42	2.56	2.70	2.84	2.98	3.20	3.54	3.68	3.97	4.39
τ/T_e	0.1	0.2	0.3	0.4	0.5	0.6	0.7	0.8	0.9	1	1.2	1.4	1.5	1.7	2

3.2.1.2 Selection of Optimisation Method

Given that the PID controller has only three tuning parameters and, consequently, the optimal tuning rules are also defined through only three parameters (coefficients), we will have a low-order optimisation problem. This is often referred to as the *nonlinear programming* problem because in most cases it is solved through application of algorithms and computer programming. Respective functions are available in commercial mathematical packages. For example, function "fminsearch" is available in MATLAB and can be used for finding optimal tuning rules.

Normally, algorithms used for unconstrained optimisation are more straightforward than those used for constrained optimisation. While the former can be relatively simple: the by-coordinate descent or gradient descent, the latter have to include some modifications through inclusion of penalties to account for the presence of constraints. In our case, there is an equality constraint that comes from the requirement of either gain or phase margin. Because of the simplicity of those formulas and the possibility of expressing one coefficient through the other two, the presence of the constraint makes the optimisation problem simpler. In fact, it allows us to reduce the order of the optimisation problem by one and look for optimal values of only two coefficients. Moreover, in the design of optimal tuning rules for the PI controller the problem becomes one-dimensional, and can thus be solved with the use of very simple algorithms. This even overcomes the problem of identifying the global minimum of the cost function. Main features of the optimisation algorithm can be shown as elements of the diagram in Fig. 3.7.

3.2.2 Tuning of Flow Loops

Now we consider a few typical processes, which often serve as elements or building blocks of a more complex process and which are often tuned separately of each other even if they interact (at least initially—before the degree of interaction with other elementary processes has been established). Among them the *flow loop* is of particular interest. This is primarily because the flow process is the most basic process in the process control industry. Flow control is used in process industries either alone or as a basic block in combination with other types of process control loops, being the secondary loop in a cascade connection. Other types of processes may include the flow process. In this case the flow controller is the secondary controller in a cascade controller connection. However, the flow controller can never be a primary controller with another controller being the secondary controller (excluding

Fig. 3.7 Logic diagram of optimisation algorithm

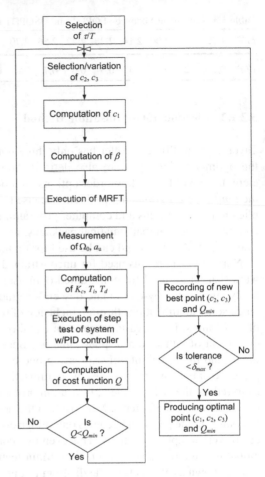

the situation involving a valve positioner, which can be considered a secondary loop with the flow controller being the primary one).

Overall the flow loop is one of the most common controls in the process industry. Analysis of the power plant distributed control system, undertaken by the author, showed that the share of the flow control loops was about 27 % of all the control loops. The flow process is also the simplest process from the point of view of the models used for identification of dynamics and controller tuning. Normally, if the valve characteristic is linear, only linear models of the flow process are used for identification and tuning. However, the practice of controller tuning gives numerous examples of the necessity of flow controller retuning from time to time caused by operating point change and for other reasons, some of which are still to be analysed. Strictly speaking, the linear model of the flow process does not fully reflect certain phenomena typical of this type of the process. Nonlinear models might better represent these phenomena. The nonlinear model of the flow loop is considered

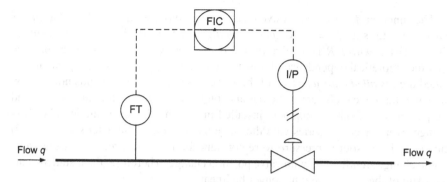

Fig. 3.8 Flow loop process diagram

in Chap. 4, in the current section, we develop only the linear model suitable to the problem being solved.

Flow loop tuning is usually done by trial and error. The reason is related to the difficulty of taking measurements of the process step response, because the response is fast, while system sampling time and especially update time of the graphic display are relatively large. This does not allow one to determine such parameters of the step response as dead time, and time required to reach 63 % of the steady state increment or the inflection point. The problem is also typical of automatic loop tuners, as most use various forms of open-loop tests, the step test being the most popular. Continuous cycling methods would be more suitable for flow loop tuning because amplitude and frequency measurements are not difficult to take. However, continuous cycling methods are more seldom used in the practice of manual tuning and auto-tuning. This is probably because practicing engineers can more easily understand step test-based methods. Therefore, for the flow loop continuous cycling methods are essentially the only choice of systematic tuning method.

A typical flow loop is given in Fig. 3.8. The notations are: FT for flow transmitter, FIC for flow controller (of PI or PID type), I/P for current-to-air pressure transducer, with air supplied to the pneumatic actuator, and the actuator driving the valve.

The formula for the flow rate q through the valve is

$$q = C_v f_v(l) \sqrt{\frac{\Delta p_v}{g_s}}, \qquad (3.25)$$

where C_v is the *valve coefficient* (often referred to as *CV of the valve*), $f_v(l) \in [0, 1]$ is a function characterising valve orifice pass area at particular *valve opening*,[1] in percent of valve CV, l is *valve opening* (or *valve lift*, which is often measured in %), Δp_v is pressure drop across the valve, and g_s is specific gravity of the fluid.

---

[1] Valve opening in this context refers to the position of the valve plug with respect to its closed position against the valve seat. It does not refer to the orifice pass area.

The function $f_v(l)$ defines valve characteristic type, with $f_v = l$ for the *linear* valve characteristic, $f_v = \sqrt{l}$ for *quick opening*, and $f_v = R^{l-1}$ for *equal percentage* characteristic, where R is a valve design parameter. Valve characteristics provide only the theoretical dependence of flow rate on valve opening; actual dependence— called the *installed characteristic* of the valve—is complex. It depends not only on the valve theoretical characteristic but also on the character of the fluid source and the parameters of other equipment installed in the line. If, for example, a heat exchanger were placed in series with the valve having linear characteristic (see [54]) then even if the source pressure were constant the pressure drop across the valve would change significantly with valve position change. Therefore, the installed characteristic of the valve would no longer be linear.

We shall assume a linear installed characteristic of the valve in the subsequent analysis. There are several reasons for such an assumption. The first is that at the design stage, the installed characteristic of the valve should be linearised by proper selection of the valve trim (valve characteristic). If, for example, the source pressure is created by a pump, which has a decreasing character of performance curve, then the selection of a valve with equal percentage characteristic would make the installed characteristic much closer to linear than would a valve having a linear characteristic. The second reason is that we are concerned with flow controller tuning, which in many cases should work only in the vicinity of a certain operating point. In this case if the valve travel from the operating point were insignificant, the nonlinearity of the installed characteristic would not have much effect on the controller tuning. This nonlinearity simply would not be revealed. The third reason is that the nonlinearity (or residual nonlinearity) of the valve installed characteristic can be linearised through the inverse nonlinearity in the controller. Therefore, we shall consider that the flow through the valve changes linearly with valve position changes.

It is worth noting that model (3.25) provides only static dependence of the flow on the valve and process parameters, and on the valve opening. Even assuming this dependence to be linear, we still have to select a proper dynamic model of the flow process. A fairly detailed nonlinear model of the flow loop is presented in Chap. 4. In the present chapter, we aim to select a fairly adequate linear model suitable for designing optimal tuning rules. While looking for a model suitable for optimisation we should mention that some books and articles recommend the use of the first-order model for the flow process with the transfer function $W(s) = \frac{K}{Ts+1}$. To illustrate the unsuitability of this model for producing tuning rules and simultaneously emphasise some necessary features of the model, let us find the best PI controller for our model, wherein the loop performance can be assessed through the step test. One can easily see that the use of just a proportional controller with high gain gives a very good (practically perfect) step response, which becomes ideally accurate if the value of the proportional gain approaches infinity and the loop remains stable. Obviously, this cannot be true. If real life were like this tuning would not be required at all. In practical application the values of the proportional and integral gains are always limited from above by stability constraints. There is only one explanation for the outcome of this imaginary test: the model is inadequate. The first-order model can still be acceptable if the flow process is part of a more complex process and not closed by the feedback. But if the flow process is considered within the loop, more accurate models are necessary. The main requirement of this model is the existence of the point of intersection of the Nyquist plot of the process and the negative part of the real axis. One can see that even the second-order model does not satisfy this requirement. Therefore, the model must be of at least third order. However, the most convenient model is that which includes delay (dead time). Two such models—FOPDT and DSOPDT—were considered above.

The presence of delay in the model allows not only a convenient approximation but also reflects the fact that there is a real delay in signal propagation through the transducer-actuator-valve dynamics. There are in addition some real and apparent delays present in the flow loop. The *real delays* are:

- the response of the flow process to valve position change due to the position of installation of the flow sensor; this delay is usually small unless there is a significant distance between the valve and the flow sensor;
- delay in the valve response to the actuator air pressure change due to Coulomb friction and backlash;
- delay in the pneumatic tubing;
- controller scan interval and processing interval (i.e. controller execution period); due to digital realisation of controllers, controller output is produced with some delay, often evaluated as one-half the scan interval or execution period (these two parameters should ideally be equal to each other); typical values of the controller scan range from 200 ms to 1 s for flow loops.

In addition to the above-listed, there are *apparent delays*, which in fact are approximations of the lags present in the loop dynamics. Among these lags are:

- sensor signal filtering or conditioning;
- dynamics of the air flow between the I/P transducer and the actuator;
- valve slew rate;
- dynamics of gas/liquid flow in the pipe.

In addition to delays, the flow process model should include dynamics and, as was mentioned earlier, first-order dynamics are not the best choice for finding optimal tuning rules. As we found, the damped second-order dynamic model provides a more adequate response of the model in MRFT. Therefore, DSOPDT dynamics are selected as the implied model for the flow process. The solution of the problem of finding optimal tuning rules is based on the DSOPDT model.

It was also mentioned earlier that the use of the equivalent FOPDT model remains convenient for the assessment of the contribution of the delay and time constants. However, in the modified RFT the relationship between T_e of the FOPDT model and T of the DSOPDT model is more complex than what is established by equations (3.23) and (3.24), because the oscillations are excited elsewhere than at the frequency point ω_π. The value of the MRFT parameter β is involved in the relationship between T_e and T, and this relationship is, therefore, unknown a priori but should be found in the process of optimisation. For this reason it is easier to start from the values of T and later, after finding an optimal point, compute the corresponding value of T_e.

We consider that the flow loop can be used in two modes of operation: as a servo system when the flow should follow the varying set point, and as a regulator, in which the set point is constant and the aim of the controller is compensation for the disturbances. We thus produce two different sets of tuning rules. If the operating mode is servo mode then usually the task of disturbance compensation remains valid, too. However, the magnitude of these disturbances may not be significant enough to adjust the tuning rules in favour of the regulator-based rules.

As mentioned above, the first-order model is unsuitable for the task of obtaining optimal tuning rules and the use of the delay in the model is essential. This issue can be traced even to Ziegler and Nichols work, when they introduced the *S-curve* as a part of their open-loop method. The use of the *S-curve* assumes, in fact, the FOPDT model. However, the FOPDT model is not very compatible with the DF method, which is used for building the model of oscillations in the RFT or MRFT. For that reason, the use of the DSOPDT model, which was introduced above, is more suitable for the flow loop. The FOPDT model will be used only in terms of "equivalence" to the DSOPDT model to determine applicable τ/T_e ranges.

Yet DSOPDT model (3.18) still has three parameters (damping ξ is fixed) that can be varied. Therefore, the problem of finding optimal tuning rules can hardly be solved with this model due to the large number of possible combinations of these three parameters, which need to be investigated. We apply a *normalisation* to model (3.18) to reduce the number of free parameters. Introduce the normalised Laplace variable $s' = Ts$, which corresponds to the following time transformation: $t' = t/T$. Noting also that the process gain K_p can be assumed as unity and accounted for later because the loop gain is determined by the product of the controller proportional gain K_c and the process gain K_p, we can use the following normalised DSOPDT model of the flow process:

$$W_p(s') = \frac{e^{-\tau/Ts'}}{s'^2 + 2\xi s' + 1}. \tag{3.26}$$

Model (3.26) has only one variable parameter τ/T, which makes the situational parameter space[2] one-dimensional instead of three-dimensional and significantly simplifies the solution of the optimisation problem. Therefore, it is necessary to compute the value of τ/T and find an optimal solution for the model (3.26). It is easy to show that the same optimal solution (in terms of coefficients c_1, c_2 and c_3) found for model (3.26) also applies to model (3.18). This happens because coefficients c_1, c_2 and c_3 establish only proportionality between the frequency of the oscillations and the time constants, and between the ultimate gain (amplitude of the oscillations) and the proportional gain: the use of the normalised frequency will produce the normalised time constants, but the use of the actual frequency will produce the actual time constants. The same rule applies to the proportional gain.

---

[2] Here, of course, we talk about the space of model parameters. The space of optimisation parameters is still two-dimensional: one dimension is eliminated through the use of the equality constraints.

3.2.2.1 Flow Loop Controller Tuning for Set Point Changes

The use of model (3.26) for finding optimal tuning rules for the set point change (the servo problem) is straightforward. This mode of operation usually occurs when the flow loop is cascade-connected to some primary loop. The system in this case is linear, and for a fixed value of τ/T optimal values of coefficients c_1, c_2 and c_3 are found through application of standard optimisation algorithms in accordance with the overall algorithm Fig. 3.7. We further investigate the following range of the ratios of dead time to equivalent time constant for the FOPDT model: $\tau/T_e \in [0.1, 1.5]$, which covers most typical values of this parameter for the flow process. The ratio τ/T for the DSOPDT model can be estimated approximately using Table 3.3. The exact relationship, however, is found in the process of optimisation after the frequency of the self-excited oscillations in the test (ultimate frequency) is determined. The time constant T_e can be found from the phase balance equation

$$-\tau \Omega_0 - \arctan(T_e \Omega_0) = -\pi + \Psi, \tag{3.27}$$

which is similar to equation (3.20) that was valid for the conventional RFT. The angle Ψ in (3.27) is created by the MRFT algorithm. It is related to the parameter β of the algorithm as follows: $\Psi = \arctan \frac{\beta}{\sqrt{1-\beta^2}}$. We express T_e from (3.27) as follows:

$$T_e = \frac{1}{\Omega_0} \tan\left(\pi - \arctan \frac{\beta}{\sqrt{1-\beta^2}} - \tau \Omega_0 \right). \tag{3.28}$$

The parameter τ/T_e can be computed afterward simply by division of τ used in the DSOPDT model by the value of T_e calculated per (3.28).

To select the optimisation criterion we find optimal solutions using the four criteria presented earlier and select the most suitable one. Using algorithm of Fig. 3.7 we find optimal solutions for the PI controller (finding optimal values of c_1 and c_2) for a certain typical value of τ/T. This values does not have to be found precisely, because the results are going to be used only for comparison of the criteria of optimisation and the selection of a criterion which will further be used. However, this "typical" value still needs to be found. What is known is the range of typical values of τ/T_e in terms of the FOPDT model, which allows us to find the corresponding value of the DSOPDT model when solving the optimisation problem. Optimisation for the values of $\tau/T \in [1.2, 4]$ allows us to select the value of $\tau/T = 1.5$ as a representative value used for comparison of different criteria; it approximately (different for different criteria) corresponds to the value of $\tau/T_e = 0.3$ of the FOPDT model. Optimal solutions for this point and the use of cost functions (3.14), (3.15), (3.16) and (3.17) are presented in Table 3.4.

Unit step responses of the closed-loop systems with PI control corresponding to the optimal solutions presented in Table 3.4 are given in Figs. 3.9, 3.10, 3.11 and 3.12. Analysis of these step responses shows that the type of response does not depend much on selection of criterion as long as constraint (3.10) is included in the

Table 3.4 Optimal solutions for DSOPDT model with $\tau/T = 1.5$ and gain margin constraint $\gamma_m = 2$

Criterion	c_1	c_2	Q_{opt}	β	τ/T_e
IAE	0.444	0.308	84.8	0.643	0.320
ITAE	0.438	0.288	307.1	0.663	0.349
ISE	0.452	0.336	59.2	0.534	0.293
ITSE	0.447	0.317	108.2	0.581	0.318

Fig. 3.9 Step response of IAE-optimal with constraint $\gamma_m = 2$ PI controller; DSOPDT model parameters: $T = 1$, $\xi = 0.8$, $\tau = 1.5$, equivalent $\tau/T_e = 0.32$ of FOPDT model

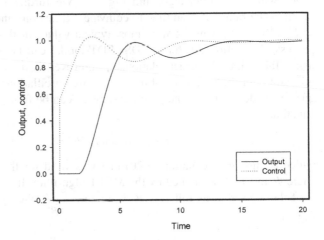

Fig. 3.10 Step response of ITAE-optimal with constraint $\gamma_m = 2$ PI controller; DSOPDT model parameters: $T = 1$, $\xi = 0.8$, $\tau = 1.5$, equivalent $\tau/T_e = 0.35$ of FOPDT model

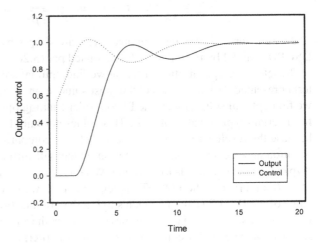

optimisation. Therefore, any criterion of the considered four would result in approximately the same performance. Thus, the use of the gain margin constraint imbedded into the MRFT algorithm provides a significant equalising effect with respect to the criterion selection effect. However, we select the ISE criterion that provides slightly higher values of c_2 than would other criteria, which better corresponds to the current practice of flow loop tuning. We proceed with the use of this criterion.

Fig. 3.11 Step response of
ISE-optimal with constraint
$\gamma_m = 2$ PI controller;
DSOPDT model parameters:
$T = 1$, $\xi = 0.8$, $\tau = 1.5$,
equivalent $\tau / T_e = 0.29$ of
FOPDT model

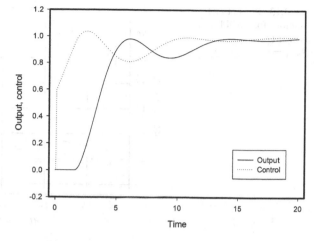

Fig. 3.12 Step response of
ITSE-optimal with constraint
$\gamma_m = 2$ PI controller;
DSOPDT model parameters:
$T = 1$, $\xi = 0.8$, $\tau = 1.5$,
equivalent $\tau / T_e = 0.32$ of
FOPDT model

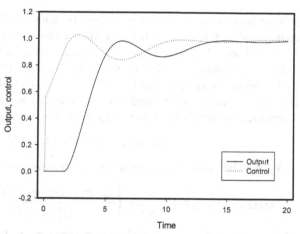

Having selected the integral criterion we then need to formulate the problem of
optimisation on the domain of *situational* (model) parameters. We have also reduced
the dimension of the model parameters to one. However, optimisation for a model
with parameter values given differs from the case in which we need to provide
optimality (in a certain sense) for the domain of *situational parameters* (possible
variation of model parameters). This aspect has been briefly considered. First, we
define the domain of *situational parameters* as the range of τ / T variations for the
DSOPDT model. For that purpose we solve the optimisation problem using the ISE
criterion for gain margin constraint $\gamma_m = 2$ and $\tau / T \in [1.2, 3.8]$ (this presumably
covers the range of interest of τ / T_e). The optimisation problem is solved separately
for each point of τ / T. The results of optimisation (parameters c_1, c_2, minimal cost
Q_{ISE}), MRFT parameter β and the equivalent τ / T_e are presented in Fig. 3.13.

It can be noted that due to the use of numeric methods aimed at speeding up the
optimisation, some values presented in Fig. 3.13 are computed with limited accuracy. This is justified by the purpose of their use: the determination of the *domain of*

Fig. 3.13 ISE-optimal solutions for DSOPDT model; $Q = Q^*_{ISE}$, $Q_{max} = \sup_{\tau/T \in D} \{Q^*_{ISE}\}$

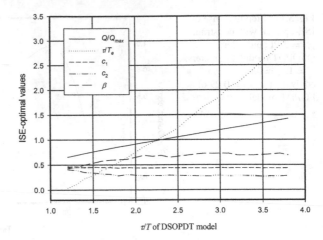

process model variations, which need not be determined very accurately due to the approximate nature of the initial data. We see from the optimisation that the *domain of process model variations* can be mapped now from the requirements for τ/T_e to the requirements to τ/T. We define the domain of the FOPDT *process model variations* as $D_{FOPDT} := \{\tau/T_e : \tau/T_e \in [0.1, 1.0]\}$ and redefine it in terms of the DSOPDT model in accordance with the established dependence (see Fig. 3.13). Therefore, our domain of *situational parameters* variations is defined as follows:

$$D := \{\tau/T : \tau/T \in [1.3, 2.3]\}. \tag{3.29}$$

We now introduce the criterion that we are going to use, which we then discuss.

$$Q_{ISE-D}(c_1, c_2, c_3) = \max_{\tau/T \in D} \left\{ \frac{\int_0^\infty e^2(c_1, c_2, c_3, \tau/T, t)\, dt}{Q^*_{ISE}(\tau/T)} \right\} \to \min, \tag{3.30}$$

where c_1, c_2 and c_3 are optimisation parameters, τ/T situational parameter and $Q^*_{ISE}(\tau/T)$ cost function for optimal solution of the following minimisation problem for a particular τ/T:

$$Q_{ISE}(c_1, c_2, c_3, \tau/T) = \int_0^\infty e^2(c_1, c_2, c_3, \tau/T, t)\, dt \to \min, \quad \tau/T \in D.$$

The criterion in (3.30) provides the ISE optimality condition on domain D. The criterion is of *min-max* type. The function $Q^*_{ISE}(\tau/T)$ provides ISE-optimal solutions for each point of τ/T in terms of criterion (3.15). Function $Q^*_{ISE}(\tau/T)$ is presented as a solid line in Fig. 3.13 (scaled) and used as a weight in criterion (3.30). The purpose of using the weight $\frac{1}{Q^*_{ISE}(\tau/T)}$ is to provide the possibility of comparison of optimal values at different points of τ/T. As a result, the criterion (3.30) does not depend on τ/T. The minimum possible value of the product $\frac{1}{Q^*_{ISE}(\tau/T)} \int_0^\infty e^2(\ldots, t)\, dt$ is 1, which it can attain only if the tuning rules in the particular point of τ/T are optimal in terms of criterion Q_{ISE}. The whole criterion (3.30) thus can be interpreted as providing minimal possible *deterioration of*

Fig. 3.14 ISE-optimal cost function values for DSOPDT model and $\gamma_m \in [2, 5]$

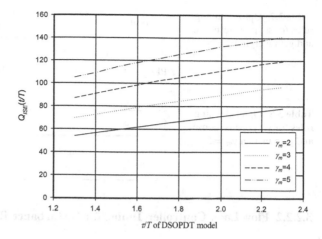

Table 3.5 Optimal tuning rules for set point response and gain margin $\gamma_m = 2$

Controller	c_1	c_2	c_3	β
P	0.500	0	0	0
PI	0.451	0.331	0	0.548

Table 3.6 Optimal tuning rules for set point response and gain margin $\gamma_m = 3$

Controller	c_1	c_2	c_3	β
P	0.333	0	0	0
PI	0.296	0.308	0	0.603

optimality on domain D due to the use of a nonoptimal solution (tuning rules) at most points (except for possibly one) of the domain D.

It is worth noting that criterion (3.30) can also be used if the domain D is *two* or *higher-dimensional*. In that case certain "meshing" of the domain is necessary to investigate function Q_{ISE} (or a different cost function) at various points of D.

The results of the solution of the optimisation problem in terms of cost function value (3.30) are presented in Fig. 3.14. These cost functions can now be used as reciprocal weights for the optimisation on the domain D—in accordance with criterion (3.30). The results of the solution of these optimisation problems for each gain margin value, in terms of coefficients c_1 and c_2 are presented in Tables 3.5, 3.6, 3.7 and 3.8. Only PI tuning rules are presented in these tables, which corresponds to the current practice. It is worth noting that the performance deterioration on the domain compared to the performance at a particular point is insignificant. For all considered cases, this deterioration is less than 1 % in terms of the cost function increase in any point of the domain D.

Table 3.7 Optimal tuning rules for set point response and gain margin $\gamma_m = 4$

Controller	c_1	c_2	c_3	β
P	0.250	0	0	0
PI	0.222	0.307	0	0.607

Table 3.8 Optimal tuning rules for set point response and gain margin $\gamma_m = 5$

Controller	c_1	c_2	c_3	β
P	0.200	0	0	0
PI	0.177	0.299	0	0.628

3.2.2.2 Flow Loop Controller Tuning for Disturbance Rejection

For optimisation based on consideration of a disturbance effect we have to consider the nature of the disturbances in the flow loop. As shown below, a mere application of an external signal to any point of the flow loop does not reflect the actual situation. Analysing equation (3.25) we see that with valve opening and specific gravity held constant the disturbance may only come in the form a change of the pressure drop across the valve due to the change of upstream (source) pressure or downstream pressure. Both these changes act the same way: an increase of upstream pressure or decrease of downstream pressure would cause an increase of pressure drop (differential) across the valve, which in turn would result in a flow increase, so that the controller would have to make adjustments to decrease the flow and bring it to the set point. And conversely, a decrease of upstream or increase of downstream pressures would result in a decrease of pressure drop and require controller action aimed at increasing the flow.

To consider this problem in more detail, we use the diagram of Fig. 3.15. In this setup the flow loop is supposed to control the flow in the line, which is accomplished through the use of the flow transmitter, the flow controller and the valve. At the same time operation of the second valve creates disturbances affecting the considered flow loop. Assume that at some point a steady state is maintained and the first valve is open at 30 % and the second valve is fully closed. We also assume that the valves are identical and that the downstream pressures for both valves are constant (due to opening to the atmosphere, for example). Flow and pressure in this steady state can be found as shown in Fig. 3.16. The operating point providing steady state flow and pressure is found at the point of intersection of the pump performance curve 1 and the system curve 2 corresponding to 30 % opening of the first valve and fully closed second valve. Therefore, the flow through the valve at this time is q_1. Assume further that the second valve opens to exactly the same 30 % at some point. The new steady state can be found at the point of intersection of the pump performance curve 1 and the new system curve, which now corresponds to both valves open at 30 %. Because both valves are identical and have the same opening the total flow is calculated as double that for one valve and so system curve 3 is created by expanding curve 2 in the horizontal direction by two times. Total flow q_3 is found at the point

Fig. 3.15 Flow loop
disturbed by valve connected
to the same source

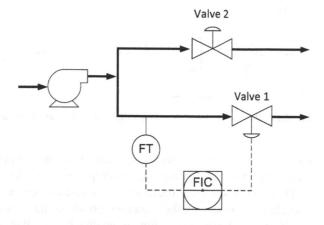

Fig. 3.16 Pressure and flow
in the system with two valves
and a pump

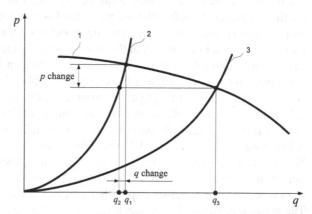

of intersection of curves 1 and 3. Yet the flow through the first valve is only half the
total flow and denoted as q_2. Therefore, we can see from Fig. 3.16 that opening of
the second valve creates a disturbance to the flow control loop being considered, in
the form of a change of pressure drop across the valve, and that if the opening of
the second valve happens quickly (so the flow loop does not have enough time to
adjust the position of the first valve) then immediately after disturbance occurrence,
the flow through the valve will be lower. The flow controller must make adjustments
in the direction of the valve opening to bring the flow to the set point.

The disturbance to the flow loop is shown in Fig. 3.17. One can see that the
disturbance is not *additive* but *multiplicative*. In fact, increase of the upstream pressure increases the loop gain, and decrease of the pressure decreases the gain. This
important feature on the one hand allows us to emphasise the importance of the
operating conditions for flow loop tuning, and on the other hand leads to the formulation of a criterion of optimisation for finding coefficients c_1, c_2 and c_3. The
following conclusion about tuning the flow loop is that tuning should be done with
process conditions that provide maximal pressure drop across the flow valve. Alternatively, if such conditions cannot be produced at the time of loop tuning then lower

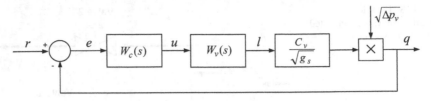

Fig. 3.17 Block diagram of flow loop disturbed by pressure drop change

gain margins should be selected, thus compensating for possible loop gain increase caused by an increase in the pressure drop across the valve.

Despite the nonlinear character of the disturbances a linear model can still be used. Let us assume that the pressure drop across the valve changes stepwise at time $t = 0$. We shall refer to the time immediately after this change as $t = 0+$ and to the time immediately before the change as $t = 0-$. Obviously, because the pressure drop change is stepwise, in the dynamic model used for tuning rule optimisation we can consider a linear model of the process with parameters corresponding to the new operating point (new steady state), which is given in Fig. 3.18. In this diagram, Δp_{0v} is the pressure drop across the valve after application of the disturbance, i.e., at $t = 0+$. It is worth noting that the variables in Fig. 3.17 are actual process variables, not increments from the steady state. This fact is important when we consider transformation of Fig. 3.17 into Fig. 3.18. However, because the model of Fig. 3.18 is linear we can use it as a model in deviations, but the deviations themselves must be from the new steady state (new operating point). Assume that the initial condition of the system at time $t = 0+$ is described by variables u_0, l_0, q_0 (we disregard for a moment other variables implicitly contained in the process transfer function). The initial conditions do not match the ones of the new steady state (at $t = 0+$). Therefore, the initial conditions of the deviations are $\Delta q(0+) \neq 0$, $\Delta l(0+) \neq 0$, and $\Delta u(0+) \neq 0$. There is, however, a relationship between the initial values of $\Delta l(0+)$ and $\Delta u(0+)$ provided by the static gain (due to the assumption that the system was in a steady state before application of the disturbance) and between $\Delta u(0+)$ and $\Delta q(0+)$ provided by the coefficients. We should also note that the value of the set point r did not change. Now using the substitution $u' = u - u(0+)$, $l' = l - l(0+)$, $q' = q - l(0+)C_v\sqrt{\frac{\Delta p_{0v}}{g_s}}$, and $r' = r - l(0+)C_v\sqrt{\frac{\Delta p_{0v}}{g_s}}$, we obtain a new equivalent system in deviations $\Delta q'(0+) = 0$, $\Delta l'(0+) = 0$, $\Delta u'(0+) = 0$ and $\Delta r'(0+) = -l(0+)C_v\sqrt{\frac{\Delta p_{0v}}{g_s}}$. From this, one can see that application of a stepwise multiplicative disturbance in the form of the pressure drop change can be equivalently analysed as a step change in the set point using the linear model. This is subject to a consideration of the pressure drop across the valve as one in the new operating point. Therefore, the optimal tuning rules obtained above for the step point change (Tables 3.5, 3.6, 3.7, and 3.8) are fully applicable to the case of tuning for the best disturbance rejection.

The approach to the system transformation considered above is valid, of course, if we assume that the dependence of flow on valve position and on pressure drop

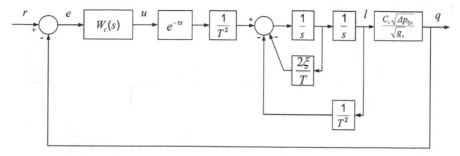

Fig. 3.18 Equivalent block diagram of flow loop disturbed by pressure drop change

is purely static. This is an accurate assumption for incompressible fluids. For compressible fluids, there is dynamic dependence of the flow on the valve opening and pressure drop, so that optimal tuning rules aimed at the best disturbance rejection should differ from those aimed at the set point following. However, the MRFT features an *equalising effect* that is revealed as very insignificant difference between the optimal values of cost functions in optimisation over the domain and optimisation in a point. As a result, one could legitimately expect that optimal tuning rules for the disturbance tuning should not be much different from those for the set point following—even if a precise flow process model were used. With the flow process considered we can now conveniently proceed with analysis of the liquid level process.

3.2.3 Tuning of Level Loops

Liquid level control in various tanks and vessels is one of the most common controls in the process industry. It can usually be categorised as (a) process control in which maintaining the level to a certain set point is the primary objective (in boiler steam drums, bottom-product and reflux drums of distillation columns, natural circulation evaporators—to name a few); this is the so-called "tight" control; and (b) the control in which large level fluctuations are allowed and even assumed—the case of the so-called surge vessels, which accumulate feedstock from one or more sources and deliver a smooth feed rate; this is also known as "averaging" control. In the latter case the primary control objective is the outflow stabilisation. In the present section, only the first category of level control objective will be considered.

Normally level is controlled by a PI or PID controller. PID controllers are used more seldom for the considered process than PI controllers because performance improvement due to introduction of the derivative term is marginal, while the derivative term itself would amplify measurement noise. In many situations satisfactory performance can hardly be achieved. This happens due to the fact that level process is an integrating process, which in combination with the integral term of the PI/PID controller results in a double integrator in the loop. The presence of the controller

Fig. 3.19 Level loop process
diagram

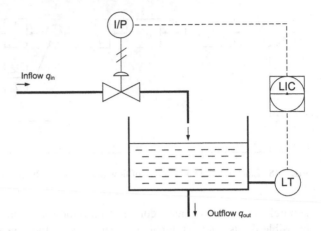

integral term is absolutely necessary to ensure zero error in a steady state, and the use of a PI controller in combination with an integrating process usually results in oscillatory transients having low damping. In summary, level process is not as easy to control in terms of providing a good performance as it might seem.

The model of the level process can be schematically represented by a tank, which has a controlled inflow and uncontrolled outflow (Fig. 3.19). In many cases the actual arrangement is the opposite: the inflow is uncontrolled and level control is done via manipulating the outflow. Yet, the second situation can be transformed into the first as shown below.

Let us assume that inflow can manipulated through some linear dynamics, so that in a steady state the inflow is proportional to the controller command. As we found above, the valve opening can be considered proportional to the controller command, but the flow is not necessarily proportional to it; plus it depends on the pressure drop across the valve. However, we disregard this dependence for the same reasons we did in the analysis of the flow controller. Besides, it can be linearised by use of the flow controller cascaded with the level controller. We can, however, note that the change in pressure drop across the control valve results in both a disturbance and a change of the loop gain. This observation leads us to the conclusion that the character of robustness that needs to be ensured in the level loop is the gain margin type of robustness.

We now write the basic equation of the process:

$$\dot{y} = \frac{1}{a}(q_{in} - q_{out}), \qquad (3.31)$$

where y is the level value, q_{in} is the controlled flow to the tank, q_{out} is the uncontrolled flow from the tank, a is the cross-sectional area of the tank (it is assumed the tank has such geometry that a is constant). If the arrangement of the flow process involves uncontrolled inflow and controlled outflow, equation (3.31) remains valid, in which q_{out} would be control and q_{in} disturbance. However, the signs of the controller gains should be changed: the higher level the larger the valve opening

Fig. 3.20 Phase portrait of PI controlled level process (underdamped)

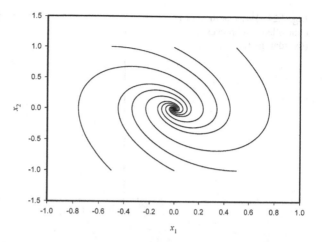

should be to stabilise the process (unlike in the process given in Fig. 3.19). The controller gain sign changes from positive to negative, cancelling the negative sign of the control term in equation (3.31), so the overall equations of the loop remain the same.

To get a better understanding of the process character and what the model requires if we are to find optimal tuning rules, we shall first consider a simplified model of the process loop. Let the process be controlled by a PI controller given by the following equation in the Laplace domain:

$$u(s) = K\left(1 + \frac{1}{Ts}\right)e(s),\tag{3.32}$$

where u is control, K is the controller proportional gain, T is the controller integral time constant, s is the Laplace variable and e is the error (the difference between the level set point and the actual level value).

At this point, let us consider that the control u produced by the controller is equal to the inflow (i.e., there are no actuator-valve dynamics): $q_{in} = u$, that the outflow is zero, and that the set point value is zero, so that $e = -y$. Rewrite equations (3.31) and (3.32) in the canonical form using the state variables $x_1 = y$ and $x_2 = \dot{y}$:

$$\begin{cases} \dot{x}_1 = x_2, \\ \dot{x}_2 = -\frac{K}{Ta}x_1 - \frac{K}{a}x_2. \end{cases}\tag{3.33}$$

Eliminate time in (3.33) by dividing the second equation by the first, and obtain the equation of the state trajectories as:

$$\frac{dx_2}{dx_1} = -\frac{K}{a}\left(\frac{1}{T}\frac{x_1}{x_2} + 1\right).\tag{3.34}$$

Depending on the parameters K and T of the controller, equation (3.34) can represent either an underdamped (oscillatory) process (Fig. 3.20) or an overdamped process (Fig. 3.21), with the origin being a focus or node, respectively.

Fig. 3.21 Phase portrait of PI controlled level process (overdamped)

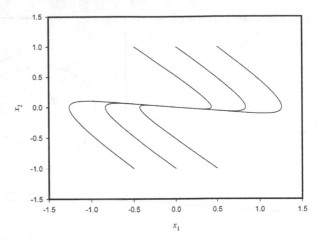

To analyse the advantages and drawbacks of each of the presented controllers with respect to the level control process, let us define the control objectives. First, a level controller is a regulator: the set point is usually constant; the main task of the controller is to attenuate (reject) possible disturbances. Second, the main disturbance applied to the level loop is the change of outflow.[3] This change is often abrupt due to consumer connection or disconnection. If, for example, the initial state is the equilibrium point (i.e., inflow is equal to outflow) then a decrease of outflow would cause an instantaneous change of the state from $\mathbf{x} = (0,0)^T$ to $\mathbf{x} = (0, x_{02})^T$. The same would happen if outflow increased. Therefore, the typical situation that needs analysing is the motion from point $\mathbf{x} = (0, x_{02})^T$. The respective system trajectories for overdamped and underdamped processes are shown in Fig. 3.22. Third, the control objective is to minimise the effect of this disturbance, which is manifested as a level increase (decrease) from the set point. The maximum level deviation corresponds to the distance between the origin and the point of intersection of the horizontal axis with the trajectory (Fig. 3.22). And fourth, one more control objective is to ensure a smooth and possibly nonoscillatory (overdamped) transient.

Let us analyse how these objectives (our criteria) are related to the controller and process parameters. The closed-loop system is a linear second-order system with the following characteristic polynomial equation:

$$P(\lambda) = \lambda^2 + \frac{K}{a}\lambda + \frac{K}{Ta} = 0, \tag{3.35}$$

which has roots:

$$\lambda_1 = -\frac{K}{2a}\left(1 + \sqrt{1 - \frac{4a}{KT}}\right)$$

---

[3] Another possible disturbance is change of pressure drop across the control valve. If significant, it can be easily mitigated by the use of the level-to-flow cascade control. In most cases it has a smaller effect on loop dynamics than outflow change.

Fig. 3.22 Phase trajectories for instantaneous outflow decrease

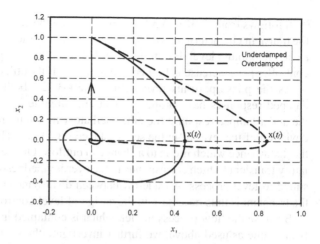

and

$$\lambda_2 = -\frac{K}{2a}\left(1 - \sqrt{1 - \frac{4a}{KT}}\right).$$

The damping ratio of the closed-loop system is, therefore, $\xi = \sqrt{\frac{KT}{4a}}$. Let us also analyse how each point $\mathbf{x} = (0, x_{02})^T$ maps to the horizontal axis, vis-à-vis the distance between the origin and the point of intersection of the trajectory, starting at $\mathbf{x} = (0, x_{02})^T$, with the horizontal axis. The analysis requires that we solve equations (3.33).

Consider the case of $\lambda_1 \neq \lambda_2$. The analytical solution can be given as follows:

$$y = c_1 e^{\lambda_1 t} + c_2 e^{\lambda_2 t}, \tag{3.36}$$

where

$$c_1 = -c_2 = -\frac{a}{K}\frac{x_{02}}{\sqrt{1 - \frac{4a}{TK}}}, \quad x_{02} = \dot{x}(0),$$

and λ_1 and λ_2 are as before. The powers of the exponents are complex in the case of the underdamped process and real in the case of the overdamped process. Let us denote by t_f the time corresponding to the intersection of the horizontal axis by the trajectory. This time can be found as a solution of the equation

$$\dot{y}(t_f) = c_1\left(\lambda_1 e^{\lambda_1 t_f} - e^{\lambda_2 t_f}\right) = 0, \tag{3.37}$$

from which time t_f can be derived as

$$t_f = \frac{1}{\lambda_2 - \lambda_1}\ln\left(\frac{\lambda_1}{\lambda_2}\right), \tag{3.38}$$

and consequently

$$y(t_f) = c_1\left(e^{\lambda_1 t_f} - e^{\lambda_2 t_f}\right). \tag{3.39}$$

It follows from (3.36)–(3.39) that by changing the controller gain K one can vary the transient characteristics of the process, including the possibility of reducing time t_f and maximum level deviation $x(t_f)$. However, first, in practice the possibility of controller gain increase is limited and normally fully utilised.

As the phase portraits reveal (Figs. 3.20–3.22), both overdamped and underdamped responses have properties that are valuable in level control. Of the two, the underdamped response (i.e., the oscillatory transient) has the smaller maximum deviation of the level from the set point (see Fig. 3.22). This deviation is caused by the disturbance which is due to a change in outflow. On the other hand, the nonoscillatory transient (which corresponds to the overdamped response) is, in general, more suited to level process. A trade-off between disturbance attenuation and quality of the transient is thus necessary in the design of level control.

Because the flow process model, which is contained in the level process model, is the same as used above, we further investigate the same range of ratios of dead time to equivalent time constant for the FOPDT model: $\tau / T_e \in [0.1, 1.5]$, which covers the most typical values for the flow process. The ratio τ / T for the DSOPDT model is found in the process of optimisation after the frequency of the self-excited oscillations in the test (i.e., ultimate frequency) is determined. The time constant T_e can be found from the phase balance equation, which is now different from the one used in the optimisation of the flow controller tuning rules (note that the integrator available in the model of the loop introduces phase lag $\pi/2$):

$$-\tau \Omega_0 - \arctan(T_e \Omega_0) = -\frac{\pi}{2} + \Psi. \tag{3.40}$$

Again, the angle Ψ in (3.40) is the one created by the MRFT algorithm. Its relationship with parameter β of the algorithm is: $\Psi = \arctan \frac{\beta}{\sqrt{1-\beta^2}}$. We express T_e from (3.40) as follows:

$$T_e = \frac{1}{\Omega_0} \tan\left(\frac{\pi}{2} - \arctan \frac{\beta}{\sqrt{1-\beta^2}} - \tau \Omega_0 \right). \tag{3.41}$$

The parameter τ / T_e can be computed from the value of τ as used in the DSOPDT model and the value of T_e as calculated per (3.41).

To select the optimisation criterion we again find optimal solutions, using the four criteria presented earlier and selecting the most suitable one. For each criterion, using the algorithm of Fig. 3.7 we find optimal solutions for the PI controller (optimal values of c_1 and c_2) for a certain typical value of τ / T. This value is found as follows. What is known is the range of typical values of τ / T_e in terms of the FOPDT model, which lets us find the corresponding value of the DSOPDT model while solving the optimisation problem. Optimisation for the values of $\tau / T \in [1.2, 4]$ allows us to select $\tau / T = 0.8$ as a representative value for the comparison of different criteria, one that approximately (being different for different criteria) corresponds to the value of $\tau / T_e = 0.3$ of the FOPDT model. Optimal solutions for this point are presented in Table 3.9.

Responses of the closed-loop systems with PI control to step change in outflow, for optimal solutions, as presented in Table 3.9 are given in Figs. 3.23 and 3.24.

Table 3.9 Optimal solutions for DSOPDT plus integrator model with $\tau/T = 0.8$ and gain margin constraint $\gamma_m = 3$

Criterion	c_1	c_2	Q_{opt}	β	τ/T_e
IAE	0.333	∞	183.6	0	0.285
ITAE	0.330	1.077	2906	0.149	0.316
ISE	0.333	∞	167.5	0	0.285
ITSE	0.333	∞	581.8	0	0.285

Fig. 3.23 Response to outflow step change of IAE/ISE/ITSE-optimal with constraint $\gamma_m = 3$ PI controller; DSOPDT model parameters: $T = 1$, $\xi = 0.8$, $\tau = 0.8$, equivalent $\tau/T_e = 0.29$ of FOPDT model

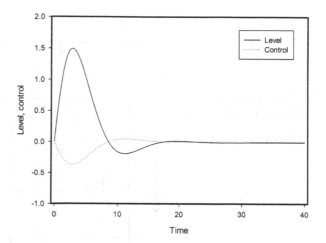

Analysis of these step responses and the results given in the table shows that three of the four criteria suggest using purely proportional control as the optimal solution. Indeed, proportional-only control is sometimes used for a level process. There is also a tendency for nearly purely proportional control use when the integral component provides an insignificant contribution. However, in most cases PI control is used, because it provides zero offset at both set point changes and disturbance application. One can see that the type of response does not change much with the criterion change as long as constraint (3.10) is included in the optimisation. And, because only the ITAE criterion suggests using PI control, which agrees with the practice of process control, we further proceed to find optimal tuning rules using the ITAE criterion.

Determine the range of τ/T corresponding to $\tau/T_e \in [0.1, 1.0]$. Exactly as in the case of the flow loop, we define the domain of the FOPDT *process model variations* as $D_{FOPDT} := \{\tau/T_e : \tau/T_e \in [0.1, 1.0]\}$ and redefine it in terms of the DSOPDT model.

We can see from the results of the optimisation that the *domain of process model variations* (*situational parameters*) can be mapped now from the requirements for τ/T_e to requirements for τ/T (see Fig. 3.25) as follows:

$$D := \{\tau/T : \tau/T \in [0.3, 2.1]\}. \tag{3.42}$$

Fig. 3.24 Response to
outflow step change of
ITAE-optimal with constraint
$\gamma_m = 3$ PI controller;
DSOPDT model parameters:
$T = 1, \xi = 0.8, \tau = 0.8$,
equivalent $\tau / T_e = 0.32$ of
FOPDT model

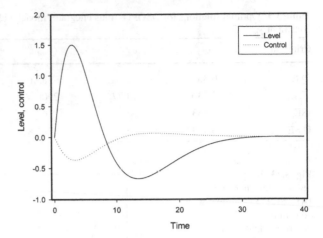

Fig. 3.25 ITAE-optimal
parameters for gain margins
$\gamma_m = 2$

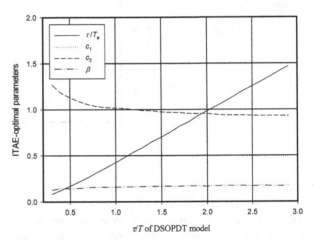

We now introduce the criterion we will use for optimisation on the domain.

$$Q_{ITAE-D}(c_1, c_2, c_3) = \max_{\tau/T \in D} \left\{ \frac{\int_0^\infty t |e(c_1, c_2, c_3, \tau/T, t)| \, dt}{Q_{ITAE}^*(\tau/T)} \right\} \to \min, \quad (3.43)$$

where c_1, c_2 and c_3 are optimisation parameters, τ/T situational parameter and
$Q_{ITAE}^*(\tau/T)$ optimal solution of the following minimisation problem for a particular τ/T:

$$Q_{ITAE}(c_1, c_2, c_3, \tau/T) = \int_0^\infty t |e(c_1, c_2, c_3, \tau/T, t)| \, dt \to \min, \quad \tau/T \in D.$$

The cost functions for optimisation on the domain are presented in Fig. 3.26.

The criterion in (3.43) provides the ITAE optimality condition on domain D. The
idea behind (3.43) is again to compensate for the effect of the model of an actual
process, so that no priority is given to situations involving faster dynamics. The
criterion is the minimum deterioration of the optimality (in terms of cost function

Fig. 3.26 ITAE-optimal
values of criteria for gain
margins $\gamma_m = 2; 3; 4; 5$

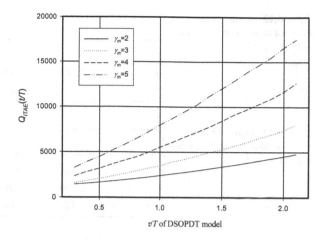

Table 3.10 Optimal tuning
rules for level controller, gain
margin $\gamma_m = 2$

Controller	c_1	c_2	c_3	β
P	0.500	0	0	0
PI	0.494	1.032	0	0.156
PID	0.468	1.344	0.080	−0.416

Table 3.11 Optimal tuning
rules for level controller, gain
margin $\gamma_m = 3$

Controller	c_1	c_2	c_3	β
P	0.333	0	0	0
PI	0.331	1.216	0	0.132
PID	0.307	1.209	0.088	−0.465

values) on the domain in comparison with optimum at the point. The results of this
optimisation are presented in Tables 3.10, 3.11, 3.12, and 3.13. It is worth noting that
unlike tuning based on a fixed ultimate frequency (corresponding to the $-\pi$ phase),
in MRFT tuning a smaller coefficient c_1 for the PID control compared to PI control
does not entail the PID controller having a smaller proportional gain, because the
frequencies of the test oscillations will be different. In fact, it is expected that the
amplitude of the oscillations for PID tuning will be smaller and the ultimate gain
higher because oscillations are excited at higher frequency (β for PID tuning is
smaller than β for PI tuning), and with decreasing magnitude frequency response,
higher frequency will result in smaller oscillation amplitude. It can also be noted
that the produced PID tuning rules are mostly of theoretical interest because it is the
PI control that is normally used in practice.

We mention that deterioration of optimality on the domain (corresponding to
the set of process models) in comparison to the optimum at a point (corresponding
to a particular process model) is about 2.7 % for gain margin $\gamma_m = 2$, 39 % for
$\gamma_m = 3$, 62 % for $\gamma_m = 4$ and 73 % for $\gamma_m = 5$ for PI tuning rules and 6 % to

Table 3.12 Optimal tuning rules for level controller, gain margin $\gamma_m = 4$

Controller	c_1	c_2	c_3	β
P	0.250	0	0	0
PI	0.249	1.484	0	0.108
PID	0.225	1.404	0.095	−0.552

Table 3.13 Optimal tuning rules for level controller, gain margin $\gamma_m = 5$

Controller	c_1	c_2	c_3	β
P	0.200	0	0	0
PI	0.199	1.703	0	0.094
PID	0.171	1.889	0.110	−0.764

9 % for PID tuning rules. This optimality deterioration is significant for higher gain margins, which makes it more profitable to select lower gain margins if the process conditions permit it (i.e., if process parameters are stable enough).

From the whole process and the results of tuning rules optimisation for the level controller, one can see that consideration of valve dynamics in the model is essential for obtaining adequate tuning rules. The provided dependencies on the value of τ/T (see Fig. 3.25, for example) illustrate this statement. Dependency on τ/T exists even within the MRFT framework, despite the equalising effect of the gain margin constraint.

The obtained tuning rules are optimal, of course, only in the sense of reaching the minimum of the ITAE cost function. This is not an absolute and other criteria can be used, too, for deriving optimal tuning rules. However, the criterion used allows us to derive rules that are optimal from the mathematical point of view, provide a good transient in most typical cases of disturbance application and agree well with the modern industrial practice of loop tuning.

3.2.4 Tuning of Pressure Loops

The pressure loop is the third most typical process in the process industry. Usually two types of pressure processes are available, which have very different dynamic properties. These are gas pressure (usually in a volume) and liquid pressure in a pipe (line). We consider both processes.

Gas pressure in a vessel can be illustrated by the diagram given in Fig. 3.27. The vessel has an inlet line with a control valve and an outlet line, with a controlled inflow and uncontrolled outflow, respectively.

Consider equations of the gas pressure process given by the diagram in Fig. 3.27. If we consider mass inflow and mass outflow we can write the equation of the gas pressure process based on the mass conservation principle.

$$\dot{m} = q_{in} - q_{out}, \tag{3.44}$$

Fig. 3.27 Pressure loop
process diagram (gas in a
vessel)

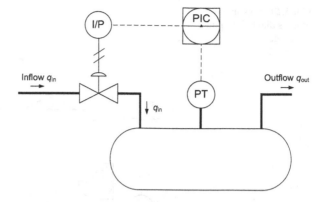

where m is the mass of the gas in the vessel, q_{in} is the controlled mass flow to the
vessel, q_{out} is the uncontrolled mass flow of the gas from the vessel.

Using the ideal gas equation

$$pV = \frac{m}{\mu}RT, \tag{3.45}$$

where p is gas pressure in the vessel, V is the vessel volume, T is the gas temperature, R is the universal gas constant ($R = 8.314 \, \text{J K}^{-1} \, \text{mol}^{-1}$) and μ is the molar
mass of the gas, we can find the time derivative of the mass of gas contained in the
vessel as follows:

$$\dot{m} = \frac{\dot{p}V\mu}{RT}. \tag{3.46}$$

In equation (3.46), we consider the process being isothermal ($T = \text{const}$; in most
cases we can assume so, at least approximately) and the volume of the vessel being
constant. We now rewrite the original mass conservation equation (3.44) as follows:

$$\dot{p} = \frac{1}{a}(q_{in} - q_{out}), \tag{3.47}$$

where $a = \frac{V\mu}{RT}$.

This is an equation of an integrating process, the dynamics of which are very
similar to the dynamics of level process (3.31). In equation (3.47), inflow can be
considered a manipulated variable and outflow a disturbance. The manipulated variable is produced by the controller command propagated through the actuator-valve
dynamics. The valves used for gas pressure control are usually the same as flow
valves (with respect to the former's dynamic response); thus, we conclude that the
dynamics of gas pressure in a vessel process are the same as those of the liquid level
process. So, we need no tuning rules other than those used in the latter process.
In fact, all tuning rules presented in Tables 3.10–3.13 can be used for gas pressure
controller tuning.

Another type of pressure process is liquid pressure in a line downstream of the
pressure source (pump), which is illustrated by the diagram in Fig. 3.28.

This type of control is normally used in situations when the hydraulic resistance
of the downstream process is constant and the maintenance of constant pressure

Fig. 3.28 Pressure loop process diagram (liquid discharge)

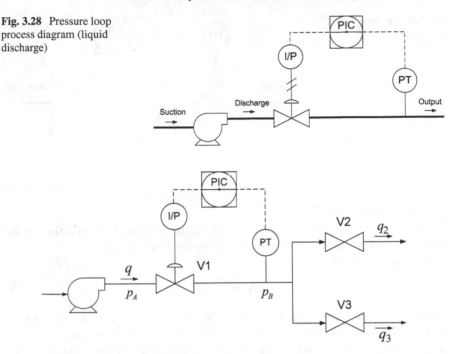

Fig. 3.29 Three-orifice arrangement process diagram

would provide stable operating conditions of the equipment. Disturbances in this control loop are caused by various consumers of the same liquid supply (i.e., pump) switching on and off—as shown in Fig. 3.29. There are other disturbances that affect the pressure of the source (pump discharge pressure) such as suction pressure change, discharge pressure change due to the change of the speed of the pump and switching on and off the pumps in a multi-pump pressure source. The character of the disturbance is the same regardless of whether it is applied at the *source* of the pressure and revealed as a change of discharge pressure, or it is at the *consumer* side due to hydraulic resistance change. Both result in change of flow and of pressure after the control valve. It seems more convenient for analysis if we present the disturbance at the consumer side, as illustrated by the diagram in Fig. 3.30.

Consider the disturbance generation mechanism. The task of the pressure loop in Fig. 3.29 is to stabilise pressure p_B by modulating the opening of valve V1 under the conditions of switching on and off consumers simulated by valve V3. Valve V2 represents the downstream process equipment, which has certain hydraulic resistance. For simplicity we assume that the individual system curve (hydraulic resistances) for each of the three valves is the same and given by curve 2 in Fig. 3.30. In the initial state (we call it mode I) let valve V3 be closed and valve V1 have opening corresponding to system curve 2. Because there are two orifices V1 and V2 connected in series in mode I, the system curve 3 is produced from curve 2 by the multiplication of curve 2's ordinate values by two. The discharge pressure of

Fig. 3.30 Disturbance generation in liquid pressure control loop

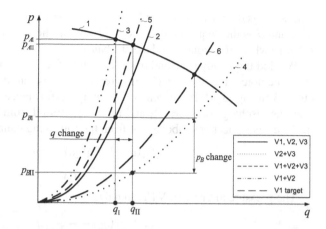

the pump p_{AI} in mode I is found at the point of intersection of pump performance curve 1 and system curve 3. The flow through valve V1 is q_I, and pressure p_{BI} is found as the pressure on system curve 2 for the equipment V2 at flow q_I. In mode II, valve V3 opens, changing the hydraulic resistance of the system. We disregard the effects of kinetic energy of the fluid such as *water hammer* despite the existence in real systems. The system curve of the parallel connection of valves V2 and V3 is found by multiplication of the abscissa values of curve 2 by two, which is shown as curve 4. Then the series connection of valve 1 and the parallel combination of V2 and V3 produces system curve 5, which is calculated by the summation of the ordinates of curves 2 and 4. The discharge pressure of the pump in mode II p_{AII} is found as a point of intersection of pump performance curve 1 and system curve 5. This point also gives the flow through valve V1, q_{II}. The pressure at the input to the equipment p_{BII} is now determined at the point on curve 4 for the flow q_{II}. The flow change due to the transition from mode I to mode II is shown as p_B change. The purpose of the control loop is, therefore, to counteract this p_B pressure change by modulating (opening) valve V1. The target opening of valve V1 is indicated as total system curve 6 (i.e., valve V1 in series with parallel connection of V2 and V3). When valve V1 arrives at the position corresponding to total system curve 6 (system curve for valve 1 can be determined as the difference between the ordinates of curve 6 and curve 4) pressure p_B again will be the same as initial.

To describe the disturbance effect mathematically, we slightly modify the example. We now consider the part of the process presented as valves V2 and V3 to be a combined orifice, which we call V'2, which can be changed. Considering that the flow is proportional to the square root of the pressure drop across the valve we can write the pressure drop as a function of the flow for valve V1,

$$f_{v1}(q) = \Delta p_{v1} = \frac{g_s}{C_{v1}^2 l^2} q^2, \tag{3.48}$$

and for valve V'2,

$$f_{v2}(q) = \Delta p_{v2} = \frac{g_s}{C_{v2}^2 d^2} q^2, \tag{3.49}$$

where p_{v2} is the pressure drop, C_{v2} is the flow coefficient for the orifice combination V'2, and d is the "equivalent" lift of V'2, so that the orifice is given by the product $C_{v2}d$ in which d is considered a disturbance.

We find the steady state and linear equations for deviations from the steady state. It can be noted that function (3.48) is represented by curve 2 and function (3.49) by curve 3 in Fig. 3.30. If we denote the pump performance curve as function $f_1(q)$ then the discharge pressure P_A and flow q are found from the point of intersection of the curves—as noted above—or from the following equations:

$$f_{v1}(q_0) + f_{v2}(q_0) = p_A, \tag{3.50}$$

$$f_1(q_0) = p_A. \tag{3.51}$$

The pressure after valve V1 is found as

$$p_B = f_1(q_0) - \frac{g_s}{C_{v1}^2 l_0^2} q^2. \tag{3.52}$$

In equations (3.50), (3.51) and (3.52), the conditions of the operating point are denoted by the subscript 0. The process gain (of the process part from valve lift to pressure) K_p can be found by differentiation of (3.52):

$$K_p = \frac{dp_B}{dl}\bigg|_{l=l_0} = \gamma \frac{dq}{dl}\bigg|_{l=l_0} + 2l_0^{-3}\frac{g_s q_0^2}{C_{v1}^2} - 2q_0 \frac{g_s}{C_{v1}^2 l_0^2}\frac{dq}{dl}\bigg|_{l=l_0}, \tag{3.53}$$

where $\gamma = \frac{df_1}{dq}\big|_{q=q_0}$ gives the slope of the pump performance curve (curve 1) at the point of the steady state.

We differentiate (3.50) with respect to l and treat it as a composite function. Then, $\frac{dq}{dl}$ is found by solving the resulting linear equation:

$$\frac{dq}{dl}\bigg|_{l=l_0} = \frac{q_0}{l_0}\frac{1}{1+\lambda}, \tag{3.54}$$

where $\lambda = \frac{C_{v1}^2 l_0^2}{C_{v2}^2 d_0^2}$ is the coefficient that represents the ratio of the orifices of valves V1 and V'2.

Finally, the formula for the process gain can be written as

$$K_p = \frac{q_0}{l_0}\frac{1}{1+\lambda}\left[\gamma + \frac{2g_s q_0}{C_{v1}^2 l_0^2}\right]. \tag{3.55}$$

In the same way, we now find the disturbance gain (from d to p_B). Differentiation of (3.52) yields

$$K_d = \frac{dp_B}{dd}\bigg|_{d=d_0} = \left(\gamma - \frac{2q_s q_0}{C_{v1}^2 l_0^2}\right)\frac{dq}{dd}\bigg|_{d=d_0}, \tag{3.56}$$

where the derivative $\frac{dq}{dd}$ is found by differentiating (3.50), treating it as we did for $\frac{dq}{dl}$, and solving the respective linear equation:

$$\frac{dq}{dd}\bigg|_{d=d_0} = \frac{q_0}{d_0}\frac{\lambda}{1+\lambda}. \tag{3.57}$$

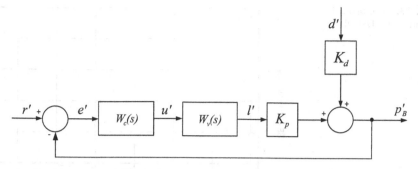

Fig. 3.31 Liquid pressure control loop

The resulting disturbance gain is

$$K_d = \frac{q_0}{d_0} \frac{\lambda}{1+\lambda} \left[\gamma - \frac{2g_s q_0}{C_{v1}^2 l_0^2} \right].$$ (3.58)

The pressure control loop can now be represented by the diagram as in Fig. 3.31. One can see from the considered example that opening of valve V3 affects the pressure loop in two ways: It exerts an external disturbance on the loop and changes the process gain. Therefore, the required gain margin needs to be assessed and optimal tuning rules for different values of gain margin have to be produced. It is also worth noting that the way the disturbance is applied to the pressure loop is similar to that of the flow loop. Therefore, the tuning rules for the pressure loop are similar to those of the flow loop. There is one difference though. In the flow loop the flow transmitter is normally located near the respective control valve. Therefore, for nearly incompressible fluids (i.e., liquid flow) we legitimately disregard the flow build-up dynamics. In pressure loops, the pressure transmitter may be located relatively far from the pressure valve, which may require inclusion of the pressure propagation delays (even for nearly incompressible fluids) in the process model. As a result, if these dynamics are accounted for as delay, this delay must be added to the delay in the actuator-valve model, that results in higher values of τ/T_e. Therefore, a slightly different range of τ/T needs to be investigated with respect to finding optimal tuning rules.

Because the model of the pressure loop is similar to that of the flow loop, we select the ISE as the criterion of optimisation. This choice was justified in the section on flow loop tuning. ISE optimisation for $\gamma_m = 3$ (considered the most typical value) shows the relationship between τ/T_e of the FOPDT model and τ/T of the DSOPDT model. Coefficients for optimal tuning rules for different τ/T are shown in Fig. 3.32 too. Considering the range of $\tau/T_E \in [0.5, 5.0]$ to be the typical range for the liquid pressure loop, we obtain the range of corresponding values of τ/T to be $[1.6, 4.6]$.

Optimal values of the ISE cost function for $\tau/T \in [1.6, 4.6]$ are presented in Fig. 3.33.

With optimal values of the cost function available we can now carry out optimisation on the domain $D := \{\tau/T : \tau/T \in [1.6, 4.6]\}$ in accordance with cri-

Fig. 3.32 ISE-optimal
solutions for DSOPDT model

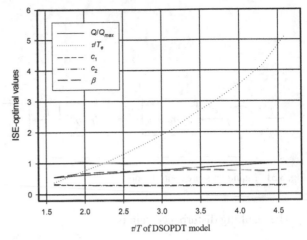

τ/T of DSOPDT model

Fig. 3.33 ISE-optimal cost
function values for DSOPDT
model and $\gamma_m \in [2, 5]$

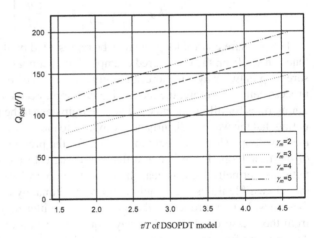

τ/T of DSOPDT model

terion (3.30). The results of the optimisation on domain D are presented in Tables 3.14, 3.15, 3.16 and 3.17. Only P and PI tuning rules are presented in these tables, which correspond to the current practice.

3.2.5 Tuning of Temperature Loops

There are a variety of different temperature loops in the process industries. Temperature control may be based on mixing of liquids or gases having different temperatures, use of heat exchanges, control of rate of chemical reactions and so on. Hence, the models of temperature processes may be significantly different. In this book we consider two types of temperature loops often found in process applications: mixing of flows having different temperatures and temperature control by manipulation of heat transfer (heat exchanges).

Table 3.14 Optimal tuning rules for pressure loop for set point response and gain margin $\gamma_m = 2$

Controller	c_1	c_2	c_3	β
P	0.500	0	0	0
PI	0.442	0.302	0	0.619

Table 3.15 Optimal tuning rules for pressure loop for set point response and gain margin $\gamma_m = 3$

Controller	c_1	c_2	c_3	β
P	0.333	0	0	0
PI	0.296	0.305	0	0.611

Table 3.16 Optimal tuning rules for pressure loop for set point response and gain margin $\gamma_m = 4$

Controller	c_1	c_2	c_3	β
P	0.250	0	0	0
PI	0.222	0.306	0	0.608

Table 3.17 Optimal tuning rules for pressure loop for set point response and gain margin $\gamma_m = 5$

Controller	c_1	c_2	c_3	β
P	0.200	0	0	0
PI	0.177	0.305	0	0.611

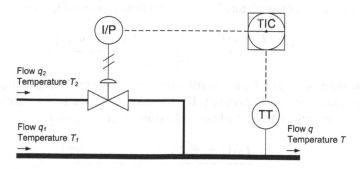

Fig. 3.34 Temperature loop process diagram (mixing of two flows)

3.2.5.1 Mixing of Flows Having Different Temperatures

A typical process of mixing two liquid flows is shown in the diagram given in Fig. 3.34. This type of process can be found, for example, in boiler steam temperature control, in which water is sprayed into the superheated steam flow producing desuperheated steam of the required temperature.

The process model is obtained by employing the laws of mass and energy conservation. Assume the temperature of the first stream T_1 is higher than of the second stream T_2, and therefore the first stream loses the amount of energy gained by the second stream. We also account for the possibility of phase change by including the energy spent on evaporation (or gained from condensation). We can now write the following energy conservation equation. We disregard here the fact that total flow is usually a controlled process variable, too, but concentrate instead on the temperature control only, thus considering a single-loop temperature control.[4]

$$q_1(T_1 - T_{mix})C_{p1} = q_2(T_{ph} - T_2)C_{p2} + q_2(T_{mix} - T_{ph})C_{p1} + q_2 L_v, \qquad (3.59)$$

where q_1 is the first stream mass flow rate, q_2 is the stream second mass flow rate, T_1 is the temperature of the first stream, T_2 is the temperature of the second stream, T_{mix} is the resulting temperature of the mixture, T_{ph} is the temperature of the phase change (evaporation, for example), C_{p1} is the specific heat capacity constant of the first stream (at constant pressure), C_{p2} is the specific heat capacity constant of the second stream (at constant pressure) and L_v is the energy required for evaporation (released at condensation) of one unit of mass. We assume the possibility of different phases (liquid and gas/vapour) in equation (3.59), so that the first stream can be a gas (as in the noted example of the steam desuperheater). We further assume that the specific heats of the liquids are different. We also take into account the fact that phase change happens at a certain temperature. So, for example, if water is sprayed into a steam stream, then at first it is heated to 100 °C and the specific heat for water needs to be applied for this part of the process; then it is evaporated, and finally it is heated to the temperature T_{mix}, so the steam specific heat constant must be applied to the last part of the process.

The temperature of the mixture T_{mix} is found from equation (3.59) as follows:

$$T_{mix} = \frac{q_1 T_1 + q_2 T_{ph} - q_2(T_{ph} - T_2)\frac{C_{p2}}{C_{p1}} - q_2\frac{L_v}{C_{p1}}}{q_1 + q_2}. \qquad (3.60)$$

We find the process gain K_p and the disturbance gain K_d by differentiating (3.60) with respect to q_2 and q_1, respectively. For the sake of simplicity, we disregard the existence of two different specific heats and assume $C_{p1} = C_{p2} = C_p$.

$$K_p = \frac{q_{10}\left(T_{ph}\left(1 - \frac{C_{p2}}{C_{p1}}\right) + T_2\frac{C_{p2}}{C_{p1}} - T_1 - \frac{L_v}{C_{p1}}\right)}{(q_{10} + q_{20})^2} \qquad (3.61)$$

and

$$K_d = \frac{q_{20}\left(T_1 - T_{ph} + (Tph - T_2)\frac{C_{p2}}{C_{p1}} + \frac{L_v}{C_{p1}}\right)}{(q_{10} + q_{20})^2}. \qquad (3.62)$$

Subscripts 10 and 20 denote the operating point about which the linearisation is done (the derivative is taken).

---

[4]If both temperature and flow have to be controlled the decoupling strategy may be used.

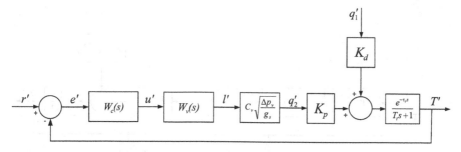

Fig. 3.35 Temperature loop model block diagram (mixing of two flows)

If no phase change is involved in the process and what is mixed is only two streams of the same liquid (hot and cold water, for example) then through setting $L_v = 0$ and $C_{p1} = C_{p2}$, equations (3.61) and (3.62) are reduced to

$$K_p = \frac{q_{10}(T_2 - T_1)}{(q_{10} + q_{20})^2} \tag{3.63}$$

and

$$K_d = \frac{q_{20}(T_1 - T_2)}{(q_{10} + q_{20})^2}. \tag{3.64}$$

It should be noted that due to the travel time of the batch of fluid from the point of injection of the second stream to the point of measurement of the mix temperature the process contains a delay τ_t that is inversely proportional to the total flow $q = q_1 + q_2$. Besides there is a process of diffusion and convection, which can be described by the first-order differential equation. The latter would add time constant T_t to the process model. Therefore, the loop model can be presented by the block diagram as in Fig. 3.35.

The servo mode (i.e., the varying reference temperature) is not typical of the temperature loop. For that reason we find tuning rules which provide optimal disturbance rejection. In the temperature loop, both temperatures T_1 and T_2 and flow q_1 can be considered disturbances. However, among these three the flow is the most important because high fluctuations of flow q_1 are usually assumed in the normal operation of the process. And it is the flow that we select to be a disturbance in finding optimal tuning rules for the temperature process. The flow loop block diagram can be transformed into the one given in Fig. 3.36.

In the diagram of Fig. 3.36, the point of the disturbance application is transposed, so that we have two delays in the loop connected in series (the second delay is contained in the transfer function $W_v(s)$). Obviously, the total delay can be considered in the model instead of two delays τ and τ_t. With this model, we can parametrise the domain D in coordinates T_t and τ_t. The time delay τ_t is determined by the transport delay and, therefore, depends on the distance between the valve and the temperature transmitter, and the fluid flow rate (translated into speed at constant pipe diameter and fluid density). This distance should be minimised. However, it cannot be made too small because proper mixing must be ensured over the time of the fluid flow

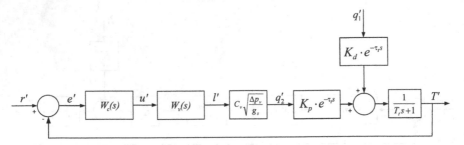

Fig. 3.36 Temperature loop model transformed block diagram (mixing of two flows)

Fig. 3.37 Optimal values of ISE cost function for temperature loop (mixing of two flows), gain margin $\gamma_m = 2$, PI controller

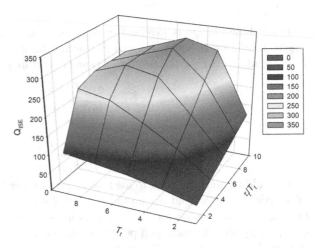

from the point of the valve connection to the point of temperature measurement. Therefore, the correct installation should ensure that $\tau_t > (3 - 5)T_t$. In practice τ_t can be even higher. Thus, for the parametrisation of domain D it would be more convenient to consider the coordinates T_t and τ_t / T_t. The solutions of the optimisation problem for every point of $D := \{(T_t, \tau_t / T_t) : T_t \in [1, 10], \tau_t / T_t \in [1, 10]\}$ for $\gamma_m = 2$ are depicted in Fig. 3.37.

ISE-optimal solutions on the domain D for different values of γ_m are presented in Tables 3.18, 3.19, 3.20 and 3.21.

3.2.5.2 Temperature Control by Manipulation of Heat Transfer

We now consider a different type of temperature process, one of heat exchange in heat exchangers, steam heated tanks, reboilers, etc. Various heat exchange processes can be modelled by a stirred tank with a constant hold-up (or a filled vessel) and a heater immersed in the tank, as illustrated by the diagram in Fig. 3.38. Cold liquid of inflow rate q_1 and temperature T_1 is supplied to the constant hold-up tank. Hot liquid with controlled flow rate q_2 and temperature T_2 is supplied to the heater, and the temperature of the outlet flow T from the tank is controlled through the

Table 3.18 Optimal tuning rules for temperature loop (mixing), gain margin $\gamma_m = 2$

Controller	c_1	c_2	c_3	β
P	0.500	0	0	0
PI	0.470	0.436	0	0.392
PID	0.490	0.289	0.125	0.208

Table 3.19 Optimal tuning rules for temperature loop (mixing), gain margin $\gamma_m = 3$

Controller	c_1	c_2	c_3	β
P	0.333	0	0	0
PI	0.313	0.436	0	0.392
PID	0.332	0.162	0.140	0.103

Table 3.20 Optimal tuning rules for temperature loop (mixing), gain margin $\gamma_m = 4$

Controller	c_1	c_2	c_3	β
P	0.250	0	0	0
PI	0.235	0.436	0	0.392
PID	0.249	0.154	0.150	0.094

Table 3.21 Optimal tuning rules for temperature loop (mixing), gain margin $\gamma_m = 5$

Controller	c_1	c_2	c_3	β
P	0.200	0	0	0
PI	0.188	0.436	0	0.392
PID	0.200	0.159	0.150	0.056

modulation of the inflow q_2 to the heater. We assume that heat is transferred from the heater to the liquid in the tank without losses and the losses are due only to the heat transfer (dissipation) between the tank and the atmosphere of temperature T_a and that temperature is uniform in the heater and in the liquid of the tank. With this model, we can write the equation of the energy conservation in the tank as follows:

$$\rho_1 V_1 \dot{T}_{out} = q_1 C_1 (T_1 - T_{out}) + h_h A_h (T_h - T_{out}) - h_d A_d (T_{out} - T_a), \quad (3.65)$$

where T_{out} is the temperature of the liquid in the tank (and the temperature of the outflow—the controlled temperature), T_1 the temperature of the inflow to the tank, T_h the temperature of the heater liquid, T_a the environment temperature, V_1 the tank hold-up volume, ρ_1 the density of the tank liquid, q_1 the mass inflow rate of the first stream, C_1 the heat capacity of the tank liquid, h_h the heat transfer coefficient between the heater and the tank liquid, h_d the heat transfer coefficient between the tank and the environment, A_h the area available for heat transfer between the heater and the tank liquid and A_d the area available for heat transfer between the tank and the environment.

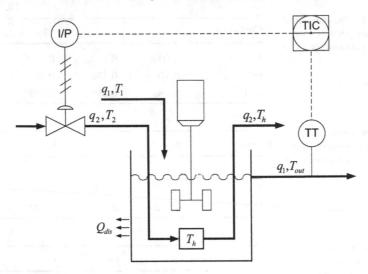

Fig. 3.38 Heat exchange temperature loop model

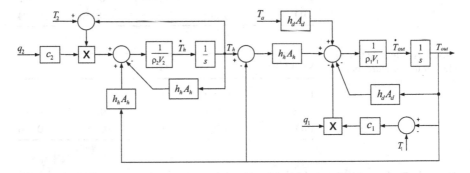

Fig. 3.39 Block diagram of heat exchange temperature loop model

We can also write the equation of the energy conservation in the heater:

$$\rho_2 V_2 \dot{T}_h = q_2 C_2 (T_2 - T_h) - h_h A_h (T_h - T_{out}), \qquad (3.66)$$

where T_2 and q_2 are the temperature and the mass flow rate for the stream entering the heater, respectively, V_2 and ρ_2 are the volume of the liquid in the heater and its density, respectively and C_2 is the heat capacity of the heater liquid.

Equations (3.65) and (3.66) are nonlinear. They can be represented by the block diagram as in Fig. 3.39. Variable q_2 is a manipulated variable, and variables T_2, T_1, q_1, T_a are disturbances. Among these, we shall consider q_1 to be the main disturbance.

Linearisation of equations (3.65) and (3.66) around the equilibrium point describing a steady mode gives the following equations for deviations from the steady state (deviations are denoted by T', q'):

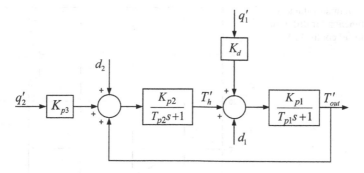

Fig. 3.40 Block diagram of heat exchange temperature loop linearised model

$$\rho_1 V_1 \dot{T}'_{out} = h_h A_h T'_h - (q_{10} C_1 - h_h A_h + h_d A_d) T'_{out}$$
$$+ C_1 (T_{10} - T_{out0}) q'_1 + q_{10} C_1 T'_1 + h_d A_d T'_a, \qquad (3.67)$$
$$\rho_2 V_2 \dot{T}'_h = C_2 (T_{20} - T_{h0}) q'_2 - (q_{20} C_2 + h_h A_h) T'_h + h_h A_h T'_{out} + q_{20} C_2 T'_2. \quad (3.68)$$

Equations (3.67) and (3.68) can be represented as the block diagram Fig. 3.40. In this diagram, the following notation for the variables is used:

$$d_1 = q_{10} C_1 T'_1 + h_d A_d T'_a,$$
$$d_2 = q_{20} C_2 T'_2,$$
$$K_d = C_1 (T_{10} - T_{out0}),$$
$$K_{p1} = \frac{h_h A_h}{q_{10} C_1 - h_h A_h + h_d A_d},$$
$$T_{p1} = \frac{\rho_1 V_1}{q_{10} C_1 - h_h A_h + h_d A_d},$$
$$K_{p2} = \frac{h_h A_h}{q_{20} C_2 + h_h A_h},$$
$$T_{p2} = \frac{\rho_2 V_2}{q_{20} C_2 + h_h A_h},$$
$$K_{p3} = \frac{C_2 (T_{20} - T_{h0})}{h_h A_h}.$$

The model presented in Fig. 3.40 is still too complex to use for parametrisation of the domain D. The fact that two disturbances, including the main one, are applied to the middle part of the model and that it has two time constants and two gains might lead one to assume the necessity of a high-dimensional and very complex parametrisation of D. We shall instead use an assumption that allows us to *simplify* the model. This assumption is valid in many cases, and even if it is not quite precise in some situations it will allow us to obtain a nearly-optimal solution because, as was shown above, the effect of the stability constraint is dominant. In many cases we can assume that the volume of the heater is much smaller than the volume of the

Fig. 3.41 Optimal values of ISE cost function for different T_p/T ratio (PI controller)

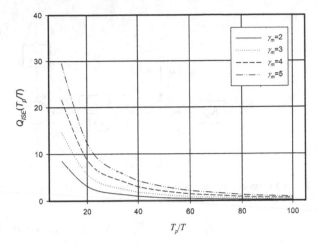

tank. In this case $T_{p2} = 0$, $K_{p2} = 1$, and the system in Fig. 3.40 transforms into the first-order system with

$$K_p = \frac{K_{p3}K_{p1}}{K_{p1} + 1} \tag{3.69}$$

and

$$T_p = \frac{T_{p1}}{K_{p1} + 1}. \tag{3.70}$$

For the temperature process, the time constant T_p is usually much higher than the time constant T of the DSOPDT model of the valve (which is also a part of the loop). As found from the results of computing, the variations of the dead time within the range of values typical of the control valve and corresponding to $\tau/T_e \in [0.1, 1.0]$ does have a significant effect on the results of the optimisation for the temperature loop. Therefore, the domain of optimisation can be defined in terms of the T_p/T ratio. Even for small size heat exchangers, the process time constant T_p is much larger than T. Therefore, we define the domain of optimisation as $D := \{T_p : T_p \in [10T, 100T]\}$ (where T is the time constant of the valve model) considering also that for $T_p/T > 100$ the results of optimisation differ insignificantly from those at $T_p/T = 100$. The results of optimisation are presented in Fig. 3.41. The curves have decreasing character, which is different from the optimal curves for other types of processes. In the current situation, the lower values of the cost function are obtained for worst-case process dynamics. This can be explained by the point of entry of the disturbance in the loop model. The disturbance enters the model at the input to the process and is thus filtered by the process. Therefore, larger time constants, which characterise slower dynamics, filter step disturbances better than smaller time constants and result in smaller deviations of the process variable.

Coefficients of ISE-optimal tuning rules for domain $D := \{T_p : T_p \in [10T, 100T]\}$ are presented in Tables 3.22–3.25 for $\gamma_m = 2, 3, 4$ and 5, respectively. One can see that optimal β is positive only for gain margin $\gamma_m = 2$ and negative

Table 3.22 Optimal tuning rules for temperature loop (heat exchange), gain margin $\gamma_m = 2$

Controller	c_1	c_2	c_3	β
P	0.500	0	0	0
PI	0.485	0.640	0	0.257
PID	0.500	0.275	0.090	0.014

Table 3.23 Optimal tuning rules for temperature loop (heat exchange), gain margin $\gamma_m = 3$

Controller	c_1	c_2	c_3	β
P	0.333	0	0	0
PI	0.325	0.704	0	0.232
PID	0.318	0.340	0.125	−0.334

in the other three cases, which implies the integrating term-dominant type of tuning for $\gamma_m = 2$ and derivative term-dominant tuning for $\gamma_m = 2, 3, 4$ (at the phase cross-over frequency of the system with the PID controller). Despite the significant difference in the PID tuning rules for the different gain margins, the rules may not necessarily produce substantially different parameters of the PID controllers (with the exception for c_1, of course). This happens because these tuning rules apply to the holistic test-and-tuning procedure and the ultimate frequency in MRFT for different gain margins is different. For example, in the PID tuning for gain margin 3 (Table 3.23), the ultimate frequency is higher than in the test for gain margin 2 because the value of β is smaller (for gain margin 3 the test is done in the second quadrant of the frequency response). Then the values of T_i and T_d for both considered cases may be very close. It can be also noted that the tuning rules for the PID controller for gain margins 4 and 5 presented in Tables 3.24 and 3.25, respectively, are mainly of theoretical interest because using higher gain margins implies that the process gain may vary, which in turn would change the gain margin of the loop.[5] As one can see, low gain margins require different tuning. As was found through simulations the use of PID tuning coefficients from Tables 3.24 and 3.25 may result in lower robustness than the use of c_2, c_3 and β from Tables 3.22, 3.23, if the process gain increases. In this case, the coefficients c_2, c_3 and β from Tables 3.22, 3.23 can be used and coefficient c_1 recalculated as inversely proportional to the specified gain margin.

3.3 Conclusions

The author's modified relay feedback test (MRFT) and the holistic approach to test and tuning are presented in this chapter. It is shown that the use of coordinated

[5]These tuning rules provide theoretical optimum; due to the possibility of process gain change they may not ensure high robustness.

Table 3.24 Optimal tuning rules for temperature loop (heat exchange), gain margin $\gamma_m = 4$

Controller	c_1	c_2	c_3	β
P	0.250	0	0	0
PI	0.244	0.729	0	0.224
PID	0.211	0.648	0.140	-0.820

Table 3.25 Optimal tuning rules for temperature loop (heat exchange), gain margin $\gamma_m = 5$

Controller	c_1	c_2	c_3	β
P	0.200	0	0	0
PI	0.195	0.722	0	0.226
PID	0.167	0.712	0.140	-0.869

test and tuning, which constitutes the holistic approach, allows one to tune PID controllers with a specified gain or phase margin.

Then, a number of typical control loops frequently encountered in the process industry are considered. Certain models of the considered process, suitable for the derivation of the tuning rules for MRFT tuning are developed. These models reflect the essential properties of the considered processes, such as integrating or self-regulating property, time constants and delays present, points of disturbance application and dependence on the operating point. All this allowed us to treat each type of the loop differently and obtain optimal tuning rules suitable for this particular process. One can see the differences between the optimal tuning rules for flow loop and level loop, for example. Even for the same values of the required gain margin, the flow loop requires more integral action, whereas the level loop tends to be nearly proportional. These observations well agree with the industrial practice of controller tuning, which confirms the validity of the approach used. Properties of the various processes considered in the chapter, which lead to the particular optimisation approach, are summarised in Table 3.26.

Properties of the MRFT such as the stability constraint used within this method helped us achieve this goal because the stability margins are satisfied in each set of tuning rules and optimality is provided for a particular process. This approach helped us to design tuning rules that are the best for a particular process subject to the condition that the process parameters satisfy some constrains or their values satisfy some conditions resulting from the definition of the domain D of possible variations of the process parameters. The definition of this domain is made using dimensionless parameters, and therefore the constraints are applied not to the process parameters but to the relationship among them. This makes the described approach suitable for the majority of practical cases because the domain is selected to cover most of situations encountered in process control. The developed optimisation on the domain allows one to obtain guaranteed results as long as the process parameters satisfy the constraints set by the definition of the domain.

Table 3.26 Disturbances and robustness type for different processes

Process	Point of disturbance input	Character of process parameter variation	Character of robustness
Flow	Output of process	Process gain	Gain margin
Level	Input to process	Process gain	Gain margin
Gas pressure	Input to process	Process gain	Gain margin
Liquid pressure	Output of process	Process gain	Gain margin
Temperature (mixing)	Inner point of process[a]	Process gain and delay	Gain margin and phase margin
Temperature (heat exchange)	Inner point of process	Process gain and time constant	Gain margin and phase margin

[a]Flow variation is considered as a factor resulting in disturbance generation and process parameter change

It was noted that the use of precise implied models of processes might ensure the development of better tuning rules. An example of such an approach to the flow loop is provided in the next chapter.

Chapter 4
Improving the Accuracy of Tuning of PID Controllers

This chapter presents a nonlinear model of the flow process controlled by a pneumatically actuated valve. Through the example of the flow loop we show that the accuracy of optimal tuning rules can be increased if precise and possibly nonlinear process models are used for optimisation. An example illustrating this approach is provided.

4.1 Improving the Accuracy of Tuning Through Nonlinear Model of Control Valve in Flow Loop

In the previous chapter, the modified relay feedback test was presented along with a number of process-specific optimal tuning rules. Among the considered processes the flow process is of particular importance because it serves as a building block for other process models. Optimal tuning rules for the flow loop were obtained using a linear model of the flow process. It was mentioned earlier though that the use of more accurate and possibly nonlinear models can potentially increase the accuracy of the tuning rules produced. This approach is applied in the present chapter to the development of the nonlinear model of the flow process and application of this model to producing tuning rules.

Besides the use of nonlinear models based on physical principles, there are other approaches to identification and tuning. Among them are such approaches as "black-box" [77] and neural networks [64, 65]. However, the availability of a nonlinear model for controller design and tuning is usually beneficial. This knowledge can allow one to use a wealth of various tuning rules and methods [6, 44, 66].

In publication [24], the MRFT was first used for *probing* in the vicinity of an operating point. In other words, it was used for assessment of the dynamic properties of the loop in the incremental sense, for different operating points and different amplitudes of the control signal (valve motion), and after that for tuning the controller on the basis of the results of this probing.

I. Boiko, *Non-parametric Tuning of PID Controllers*, Advances in Industrial Control, DOI 10.1007/978-1-4471-4465-6_4, © Springer-Verlag London 2013

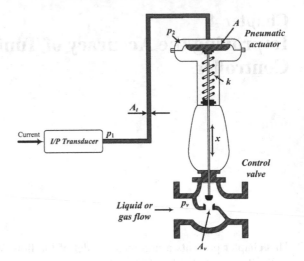

Fig. 4.1 I/P transducer, actuator and valve

In this chapter, the dynamic nonlinearity of the flow process is investigated through building a model of the pneumatically actuated valve based on fluid dynamics laws. The model still assumes the linear static dependence of the controlled flow rate on the controller command. It is shown that in spite of this linear static dependence, the dynamics of this process are strongly nonlinear, which will result in different tuning parameters of the PID controller for different amplitudes of the relay in the test and different operating conditions (different set points for the flow). Dependence of the required tuning parameters on the value of the relay amplitude in the (M)RFT and selection of the operating point are analysed.

4.1.1 Model of Flow Process

The model of the flow process is based on the mechanics of the current-to-pressure (I/P) transducer, the pneumatic actuator, the control valve and gas or liquid flow through the valve (Fig. 4.1). There also exists another arrangement for the control of the pneumatically actuated valve: one using a positioner, which involves the measurement of the valve (stem) position and a feedback to the transducer. However, in this chapter we consider only the first type of actuation arrangement.

The equation of the stem motion is

$$m\ddot{x} + b\dot{x} + kx = (p_2 - p_a)A - \Delta p_v A_v + F_f, \tag{4.1}$$

where x is the valve (stem) position, m is the combined mass of the valve stem, diaphragm and of other moving parts, b is the coefficient of viscous friction, k is the spring rate, p_2 is the pressure (absolute) above the diaphragm, p_a is the atmospheric pressure, $\Delta p_v A_v$ is the force exerted on the plug due to the pressure drop across the valve and F_f is the Coulomb friction. To isolate the cause of the effects we are going to analyse in this system, we will assume that $F_f = 0$ and $p_v A_v = 0$.

However, the addition of nonzero values of the referenced variables in the presented model would not significantly change the results of the analysis and simulations, nor would it change the resulting conclusions. We will also assume that the flow through the valve changes linearly with the valve position change (linear valve), and the dynamics of the flow build-up can be approximated by a delay, which can be added to the delay introduced by the controller. A precise model of the flow through the valve can be based on either a lumped model (similar to the one used below for the modelling of the transducer-actuator dynamics) or the Navier–Stokes equations. We use the delay approximation so we may isolate the specific problem we wish to study.

We shall assume that the I/P transducer produces output pressure p_1 (absolute) proportional to the supplied current, and that the reaction of the I/P transducer to the current change is fast enough, so that the dynamics of the I/P transducer can be described by a small time delay or even neglected, as this time delay can be accounted for in the processing delay of the control system (which usually varies from 200 ms to 1 s for flow loops). The static dependence of the transducer output pressure on the input current is linear, with 4–20 mA current range corresponding to 3–15 psi (gauge) pressure range (this is the most common industry standard). However, pressure p_2 is not equal to pressure p_1 (in transients) because there is a transmission line (tube) between the two chambers: the transducer chamber (in which the transducer pressure is sensed and used for the feedback in the transducer internal control system) and the actuator chamber. Normally pneumatic tubes are 1/4 or 3/8 OD and made of metal or plastic. We shall use the following model of the air flow between these two chambers. The mass air flow from chamber 1 to chamber 2 is given by the following formula based on the St. Venant and Wantzel formula [12]:

$$G_{12} = A_t c_d p_1 \sqrt{\frac{g\gamma}{RT_1}\left(\frac{2}{\gamma+1}\right)^{\frac{\gamma+1}{\gamma-1}}} \,\Psi\left(\frac{p_2}{p_1}\right), \qquad (4.2)$$

where A_t is the smallest cross-sectional area of the tube between the I/P transducer and the actuator, c_d is the discharge coefficient (so that $A_t c_d$ is the effective cross-sectional area of the tube), g is the gravity constant, R is the universal gas constant, T_1 is the air temperature in chamber 1, γ is the isentropic coefficient ($\gamma = 1.4$ for air) and $\Psi(p_2/p_1)$ is the flow function given by

$$\Psi\left(\frac{p_2}{p_1}\right) = \begin{cases} 1 & \text{if } \frac{p_2}{p_1} \le \beta_c, \\ \sqrt{\frac{2}{\gamma-1}\left(\frac{\gamma+1}{2}\right)^{\frac{\gamma+1}{\gamma-1}}}\sqrt{\left(\frac{p_2}{p_1}\right)^{\frac{2}{\gamma}} - \left(\frac{p_2}{p_1}\right)^{\frac{\gamma+1}{\gamma}}} & \text{if } \frac{p_2}{p_1} > \beta_c, \end{cases} \qquad (4.3)$$

where $\beta_c = \left(\frac{2}{\gamma+1}\right)^{\frac{\gamma}{\gamma-1}}$ is the critical pressure ratio ($\beta_c = 0.528$ for air). A graph of function $\Psi(p_2/p_1)$ is presented in Fig. 4.2.

We note that air may flow not only from the transducer to the actuator but in the opposite direction too, when the command to the valve is to "close" (the current supplied to the I/P transducer decreases). In this case, the air flow from the actuator to

Fig. 4.2 Graph of function
$\Psi(p_2/p_1)$

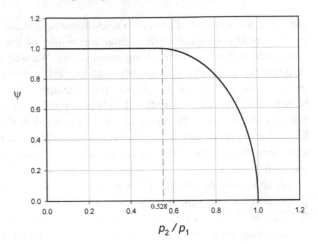

the transducer (where it is released to the atmosphere) G_{21} can be described by the same equations (4.2) and (4.3) with pressures p_1 and p_2 swapped in the formulas. Considering that both cases, $p_2 > p_1$ and $p_2 < p_1$ are possible, we want to design a nonlinear function that would encompass both these cases. However, if we assumed p_2/p_1 as the argument of this nonlinearity we would create an asymmetric nonlinearity, which would be inconvenient for analysis. It would be more convenient to deal with a symmetric nonlinearity encompassing the cases of $p_2 > p_1$, $p_2 < p_1$ and $p_2 = p_1$. With that purpose in mind, we first introduce the following nonlinear function, where instead of argument p_2/p_1 we use the logarithm of this value:

$$\Phi^+(\eta) = \begin{cases} 1 & \text{if } \eta \geq -\ln\beta_c, \\ \kappa\sqrt{e^{-\eta\frac{2}{\gamma}} - e^{-\eta\frac{\gamma+1}{\gamma}}} & \text{if } 0 \leq \eta < -\ln\beta_c, \end{cases} \quad (4.4)$$

where $\kappa = \sqrt{\frac{2}{\gamma-1}(\frac{\gamma+1}{2})^{\frac{\gamma+1}{\gamma-1}}}$ and $\eta = -\ln(p_2/p_1)$. Function $\Phi^+(\eta)$ can possess only nonnegative values; and we now have to design a function that would be odd-symmetric, using definition (4.4). We introduce the following odd-symmetric function:

$$\Phi(\eta) = \begin{cases} -1 & \text{if } \eta \leq \ln\beta_c, \\ \kappa\,\text{sign}(\eta)\sqrt{e^{-|\eta|\frac{2}{\gamma}} - e^{-|\eta|\frac{\gamma+1}{\gamma}}} & \text{if } \ln\beta_c < \eta < -\ln\beta_c, \\ 1 & \text{if } \eta \geq -\ln\beta_c. \end{cases} \quad (4.5)$$

The plot of function $\Phi(\eta)$ is presented in Fig. 4.3. Considering the introduced functions and the equation for pressure change in chamber 2 based on the ideal gas equation $p_2 V = \frac{m}{\mu}RT_2$, where $V = V_0 + Ax$ is the volume of chamber 2, with V_0

Fig. 4.3 Graph of function $\Phi(\eta)$

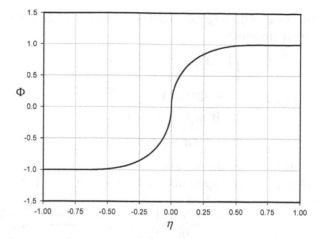

being the volume at $x = 0$ and $T_2 = T_1 = T$, write the state equations of the plant (process) as follows.

$$\dot{x} = v,$$

$$\dot{v} = \frac{1}{m}(-bv - kx + (p_2 - p_a)A),$$

$$\dot{p}_2 = \frac{1}{V_0 + Ax}\left(\frac{G_{12}}{\mu}RT - p_2Av\right) \tag{4.6}$$

$$= \frac{1}{V_0 + Ax}\left(\frac{\sqrt{RT}}{\mu}A_t c_d p_1 \kappa \sqrt{g\gamma\left(\frac{2}{\gamma+1}\right)^{\frac{\gamma+1}{\gamma-1}}}\Phi(\eta) - p_2Av\right),$$

where $\eta = -\ln(p_2/p_1)$. Equations (4.6) are, therefore, the set of three equations for the stem (valve) position, stem (valve) velocity and pressure in chamber 2. Pressure in chamber 1 is considered a control input. Equations (4.6) are nonlinear, and linearisation around the equilibrium point $p_2 = p_1$, $v = 0$, $x = x_0$ would be of much help for designing a controller from the process dynamics around the equilibrium point. This approach is explored in the following section.

4.1.2 Lyapunov Linearisation of Flow Process Dynamics

We now undertake the task of obtaining the Lyapunov linearisation of the nonlinear system around the equilibrium point corresponding to a certain flow set point value. This is not a trivial task because the nonlinear function is not Lipschitz continuous in the point $p_2 = p_1$ (it is easy to show that $\lim_{\eta \to 0} \frac{d\Phi(\eta)}{d\eta} = \infty$), and generally the Lyapunov linearisation cannot be applied to such a system. To overcome this problem we replace the original nonlinearity $\Phi(\eta)$ with the following Lipschitz continuous approximation:

$$\Gamma(\eta,\epsilon) = \begin{cases} 1 & \text{if } \eta \leq \ln \beta_c, \\ -\kappa\, \text{sign}(\eta)\left\{\sqrt{e^{-(|\eta|+\epsilon)\frac{2}{\gamma}} - e^{-(|\eta|+\epsilon)\frac{\gamma+1}{\gamma}}} - \sqrt{e^{-\epsilon\frac{2}{\gamma}} - e^{-\epsilon\frac{\gamma+1}{\gamma}}}\right\} \\ \qquad \text{if } \ln\beta_c < \eta < -\ln\beta_c, \\ -1 & \text{if } \eta \geq -\ln\beta_c. \end{cases} \quad (4.7)$$

One can see that for any $\epsilon > 0$ the following identities hold: $\lim_{\epsilon\to 0}\Gamma(\eta,\epsilon) = \Phi(\eta)$, $\Gamma(0,\epsilon) = 0$ and $\lim_{\eta\to 0}\partial\Gamma(\eta,\epsilon)/\partial\eta \neq \infty$. In fact,

$$\left.\frac{\partial\Gamma(\eta,\epsilon)}{\partial\eta}\right|_{\eta=0} = -\kappa\frac{(\gamma+1)e^{-\epsilon\frac{\gamma+1}{\gamma}} - 2e^{-\epsilon\frac{2}{\gamma}}}{2\gamma e^{-\frac{\epsilon}{\gamma}}\sqrt{1 - e^{-\epsilon\frac{\gamma-1}{\gamma}}}}, \quad (4.8)$$

which is a nonzero value at any $\epsilon > 0$.

The Lyapunov linearisation of system (4.6), in which nonlinearity $\Phi(\eta)$ is replaced with $\Gamma(\eta,\epsilon)$, results in the following model:

$$\dot{\mathbf{x}} = \mathbf{Ax} + \mathbf{B}u, \quad (4.9)$$

where $\mathbf{x} = [x\ v\ p_2]^T$, $u = p_1$,

$$\mathbf{A} = \begin{bmatrix} 0 & 1 & 0 \\ a_{21} & a_{22} & a_{23} \\ a_{31} & a_{32} & a_{33} \end{bmatrix}, \qquad \mathbf{B} = \begin{bmatrix} 0 \\ 0 \\ b_3 \end{bmatrix},$$

$$a_{21} = -\frac{k}{m},$$

$$a_{22} = -\frac{b}{m},$$

$$a_{23} = \frac{A}{m},$$

$$a_{31} = -\frac{A}{(V_0 + Ax_0)^2} \times \left(\frac{\sqrt{RT}}{\mu}A_t C_d p_{10}\kappa\sqrt{g\gamma\left(\frac{2}{\gamma+1}\right)^{\frac{\gamma+1}{\gamma-1}}}\Gamma(0,\epsilon) - p_{20}Av_0\right),$$

$$a_{32} = -\frac{p_{20}A}{V_0 + Ax_0},$$

$$a_{33} = \frac{1}{V_0 + Ax_0} \times \left(\frac{\sqrt{RT}}{\mu}A_t C_d \kappa\sqrt{g\gamma\left(\frac{2}{\gamma+1}\right)^{\frac{\gamma+1}{\gamma-1}}}\left.\frac{\partial\Gamma(\eta,\epsilon)}{\partial\eta}\right|_{\eta=0} - Av_0\right),$$

$$b_3 = \frac{1}{V_0 + Ax_0}\frac{\sqrt{RT}}{\mu}A_t C_d \kappa\sqrt{g\gamma\left(\frac{2}{\gamma+1}\right)^{\frac{\gamma+1}{\gamma-1}}} \times \left(\Gamma(0,\epsilon) + \left.\frac{\partial\Gamma(\eta,\epsilon)}{\partial\eta}\right|_{\eta=0}\right).$$

Subscript 0 on variables x, v, p_2 and p_1 is used to denote the steady state values. The transfer function from control p_1 to position x, which corresponds to linear model (4.9), is given by the following transfer function: $W(s) = \mathbf{C}(s\mathbf{I} - \mathbf{A})^{-1}\mathbf{B}$, where $\mathbf{C} = [1\ 0\ 0]$. It can be expressed through the coefficients given above as follows:

$$W(s) = \frac{a_{23}b_3}{s^3 - (a_{33} + a_{22})s^2 - (a_{21} - a_{22}a_{33} - a_{23}a_{32})s + a_{21}a_{33} - a_{23}a_{31}}.$$

(4.10)

We now find that the limit $\lim_{\epsilon \to 0} W(s)$ is $W^*(s)$, which can be derived as follows:

$$W^*(s) = \frac{a_{23}}{s^2 - a_{22}s - a_{21}} = \frac{A}{ms^2 + bs + k}.$$

(4.11)

It can be considered as the plant equation subject to p_2 being the control signal, that is $p_2 = p_1$.

One can see that the limit $\lim_{\epsilon \to 0} W(s) = W^*(s)$ results in the reduction of the third-order dynamics to second-order dynamics. Formula (4.11) is a widely used model of actuator dynamics. However, as shown above, this is an approximation of more complex third-order nonlinear dynamics, and the relation between the two is demonstrated.

We now aim to show that the complexity of the third-order dynamics should be accounted for in controller design. This is done via probing based on MRFT.

4.1.3   Local Probing of Incremental Nonlinear Dynamics Through MRFT

Because the Lyapunov linearisation fails to provide an adequate linear model in the vicinity of an equilibrium point, the problem can be posed of how local behaviour of the process can be described, so that a PID or another controller can be designed. In fact, this problem has been successfully solved in practice over the time since the 1940s when the closed-loop tuning method was proposed by Ziegler and Nichols. Because this method involves oscillations of finite amplitude, probing is applied to a finite range of the nonlinearity, and, therefore, the stability and performance properties of the system after introduction of the controller would be adequate only if the range of actuator pressure variations in operation of the system were to match the range of actuator pressure variations in the Ziegler–Nichols test [88]. Consequently, a numeric estimate of equivalent linear dynamics can be obtained, for example, through the describing function method [8] (if the controller command is close to a sinusoid signal, which is the case when method [88] is used for tuning). A similar probing is ensured by the relay feedback test [5] and its modifications. In this case, due to the controlled character of the amplitude (of the relay signal), it becomes possible for one to estimate local dynamics in a narrower or wider range of increments from the equilibrium point by changing the control amplitude.

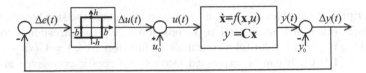

Fig. 4.4 Probing through modified relay feedback test (via increments to steady state variables)

We shall use the modified RFT presented in Chap. 3 for this kind of probing. It offers some advantages over the conventional test as it allows us to design of a PID controller with specified gain or phase margin (for linear processes) without knowledge of the process model.

Because we do not aim to restore/approximate the local model of the process (due to the measurement of just two parameters of the test, which is insufficient for building an adequate model), we will characterise the local dynamics by measuring the frequency and amplitude of oscillations in the test, as well as by the parameters of the PID controller designed with the use of some specific tuning rules. The same tuning rules will be applied to all operating points and amplitudes of the relay signal. We will thus be able to compare the required controller tuning in different operating conditions and different control amplitudes. Through this comparison, we will analyse the character of the nonlinearity in the flow process.

Doing this we can ensure that after introduction of the PID controller the system will be stable if the deviations from the operating point are within the ranges covered by the control and process variable amplitudes in the test. The logic behind this is that different amplitudes of the test will cover different ranges of the nonlinearity, and among all these measurements the most conservative values can be selected for the PID controller design (tuning). In this case, regardless of the fluctuations of the variable values (with respect to the equilibrium point) the system will remain stable.

Local probing through the MRFT is illustrated by the block diagram of Fig. 4.4. The variables of a steady state are identified by subscript 0. The MRFT is, therefore, implemented with respect to control increments, and measurements of output increments are taken.

The MRFT is realised through the algorithm given by formula (3.1) applied to the increments from the steady state values, with parameter β and coefficients defining the tuning rules selected from Tables 3.5, 3.6, 3.7 or 3.8.

4.1.4 Example of Analysis

The analysed example of the flow process involves typical parameters of the control valve and actuator with instrument air pressure in the actuator varying within 3–15 psig and valve stroke 45 mm. Other actuator and valve parameters are as follows:

- effective diaphragm surface area $A = 0.0028 \text{ m}^2$,
- effective tube surface area $A_t = 0.00001963 \text{ m}^2$,

- spring rate $k = 5200$ N/m,
- mass of stem and associated moving parts $m = 1.36$ kg,
- viscous friction $b = 2425$ kg/s,
- maximal water flow (at 100 % valve opening) 68,000 kg/h,
- and valve static characteristic is *linear*.

The results of application of the MRFT with parameter $\beta = 0.195$ and four different amplitudes to the nonlinear process model presented above are given in Tables 4.1 and 4.2. All presented results are computed for the 0.5 s dead time in the flow loop, which corresponds to the typical execution period of the DCS and some delay in water flow build-up. In particular, the values of ultimate period $2\pi/\Omega_0$ and ultimate gain (determined as $4h/(\pi a_0)$, where a_0 is oscillation amplitude) are presented for different operating points and different values of the relay amplitude. In addition, ultimate period and ultimate gain are computed for the process represented by linear model (4.11), which can be considered as a limiting case at relay amplitude $h \to 0$.

Ultimate period and ultimate gain values are presented in Tables 4.1 and 4.2, respectively. One can see that the studied nonlinear dynamics reveal a strong dependence on relay amplitude and to a lesser degree on operating point; in particular, the smaller the relay amplitude the smaller the ultimate period, which in general fully agrees with real process observations [24]. This conclusion can rarely be observed and analysed in practical applications due to the fact that tuning is usually done on a live process, which normally does not allow application of large relay amplitudes because it may disturb the process. Moreover, flow loop dynamics may be affected by other circumstances (source pressure, for example), and it is not always possible to draw definite conclusions about the effect of relay amplitude from observations of the process in industrial applications. This makes the obtained results particularly interesting and useful to the practice of process control, which may potentially influence the development of new tuning rules and possibly cause the revision of available ones. Comparison between the results of Tables 4.1 and 4.2 shows that change of relay amplitude mainly results in a change in ultimate period but not in ultimate gain. This leads us to conclude that the nonlinearity of the transducer-actuator dynamics is dynamic and could possibly be approximated by a higher time constant in the linearised model corresponding to higher values of the relay amplitude (i.e., higher amplitude of the induced oscillations).

The results of non-parametric tuning of the PI controller in accordance with sample tuning rules with $c_1 = 0.33$, $c_2 = 0.8$, for different operating points and amplitudes of the relay signal are presented in Tables 4.3 and 4.4. These results demonstrate that tuning strongly depends on the choice of the relay amplitude. The example shows that the resulting controller gains may differ by a factor of more than 1.5 for different operating points and relay amplitudes. Usually the choice of relay amplitude is limited from below by the requirement that noise and fluctuations due to pressure changes be exceeded and from above by considerations that the test itself does not disturb the process. The optimal values of the amplitude vary between 5 and 20 %, with preference given to the lower value when possible. Therefore, the fact of different operating points and relay amplitudes must be accounted for in loop

Table 4.1 Ultimate period values in modified RFT

Operating point (flow %)	Linear model (4.11)	$h = 1\,\%$	$h = 5\,\%$	$h = 10\,\%$	$h = 20\,\%$
10	1.63	1.83	2.10	2.30	
25	1.63	1.82	2.08	2.27	2.56
50	1.63	1.81	2.05	2.23	2.50
75	1.63	1.80	2.02	2.20	2.45

Table 4.2 Ultimate gain values in modified RFT

Operating point (flow %)	Linear model (4.11)	$h = 1\,\%$	$h = 5\,\%$	$h = 10\,\%$	$h = 20\,\%$
10	2.66	2.63	2.60	2.59	
25	2.66	2.63	2.60	2.59	2.61
50	2.66	2.63	2.60	2.59	2.60
75	2.66	2.63	2.61	2.59	2.60

Table 4.3 Proportional gain K_c (sample tuning rules of [27])

Operating point (flow %)	Linear model (4.11)	$h = 1\,\%$	$h = 5\,\%$	$h = 10\,\%$	$h = 20\,\%$
10	0.87	0.86	0.85	0.85	
25	0.87	0.86	0.85	0.85	0.85
50	0.87	0.86	0.85	0.85	0.85
75	0.87	0.86	0.85	0.85	0.85

Table 4.4 Integral time constant T_i (sample tuning rules of [27])

Operating point (flow %)	Linear model (4.11)	$h = 1\,\%$	$h = 5\,\%$	$h = 10\,\%$	$h = 20\,\%$
10	1.30	1.46	1.68	1.84	
25	1.30	1.45	1.66	1.82	2.05
50	1.30	1.45	1.64	1.78	2.00
75	1.30	1.44	1.62	1.76	1.96

tuning through respective adjustments, which involve increase of the gain margin to ensure loop stability and satisfactory performance. In particular, preference should be given to tuning based on higher amplitudes in the (M)RFT (to higher values of T_c from Table 4.4) as giving more conservative and safe tuning parameters for the PID controller. Therefore, if the test is implemented with smaller amplitudes of the relay then some adjustments (further increase of T_c) to the tuning rules have to be made to ensure the necessary stability margins.

4.2 Optimal Tuning Rules for Flow Loop, Based on Nonlinear Model

With the nonlinear model of the flow process available (formula (4.6)), we now approach the task of generating optimal tuning rules based on this model. Because the nonlinear model has a much higher number of parameters, and for designing the domain of optimisation D, we have to consider various combinations of a large number of parameters, the direct approach is hardly feasible. Fortunately, valve design already assumes some combinations of parameters necessary for its normal operation: Spring hardness must correspond to the force exerted by the air pressure on the diaphragm for the valve to be operable in the full range of travel; the mass of moving parts is approximately proportional to valve and actuator size; friction is related with stem diameter and is, therefore, approximately proportional to the valve size; etc. In fact, most parameters depend on valve size, so that the actuator-valve dynamics for small-size valves have much in common with the dynamics of large size valves, with the exception of the time scale. All processes happen faster in smaller-size valves. However, air flow dynamics cannot be scaled in the same way. Moreover, very often the same I/P transducers are used with small and bigger size valves. An example of such a transducer is Fisher DVC6000®️ digital valve controller[1] widely used in the process industry. Adaptation to various actuator sizes is done only through the use of two or three different sizes of the pneumatic tubes to provide larger air flow rates for larger actuators and valves.

We shall use this fact in the parametrisation of the domain D that we are going to apply. We shall suppose that such actuator-valve parameters as mass, friction, spring hardness and diaphragm area are selected in proper proportion, so that only time scale of the solution of equation (4.1) would change from one valve to another. Equation (4.1) is just a second-order differential equation, which after replacement of the Coulomb friction with some equivalent viscous friction becomes a second-order transfer function with gain and time constant not affecting the optimal values of tuning rule coefficients. In essence, we assume that equation (4.1) provides some "standard" second-order transfer functions with the time constant, which may change from one valve to another, but gain and damping ratio remain constant (in fact, we consider only damping ratio constant because gain does not have any effect on tuning rules). In that case only the parameters of equations (4.2) and (4.3) can describe the variety of actuators and valves, which realistically are just the pneumatic tube effective area $A_t c_d$. Therefore, we now carry out parametrisation of domain D through the use of this parameter.

However, this time we are not going to find one optimal solution over the whole domain but instead optimal solutions in three characteristic points of domain D corresponding to three different values of $A_t c_d$. These three different values of the tube effective area $A_t c_d$ correspond to a small valve, a medium-size valve and a large

---

[1]It provides wider functionality than a conventional I/P transducer, including positioning mode, possibility of using various pressure sources, calibration, etc.

Table 4.5 ISE-optimal tuning coefficients for PI controller	Tube size	Coefficient	Gain margin			
			2	3	4	5
	$0.5A_t$	c_1	0.43	0.26	0.17	0.14
		c_2	0.26	0.20	0.15	0.15
	A_t	c_1	0.44	0.27	0.17	0.14
		c_2	0.30	0.22	0.15	0.15
	$2A_t$	c_1	0.44	0.28	0.17	0.16
		c_2	0.30	0.25	0.15	0.21

valve, with parameter $A_t c_d$ having a larger value, then a smaller value, and then the smallest value, respectively. The small valve is thus simulated by a large value of $A_t c_d$ and vice versa. Because the time constants are scalable (the tuning rules are invariant to the time scale) we can take an example of a typical actuator and a valve and investigate the dependence of the optimal tuning rules on the parameter $A_t c_d$.

The system having the parameters presented above, in Sect. 4.1.4, was considered in paper [73]. Optimal parameters of the tuning rules for the MRFT with four different gain margins: $\gamma_m = 2, 3, 4, 5$, and for three different pneumatic tube sizes: nominal (given in Sect. 4.1.4), half the nominal, and twice the nominal, were found using the ISE criterion. More detailed analysis with the use of other criteria for optimisation is presented in [74]. Because of the dependence of the results on the choice of operating point and amplitude of the relay, the considered operating point was selected to correspond to 50 % of the valve opening. Also, 10 % of the relay amplitude in the MRFT and in the step test for the evaluation of the cost function were selected as typical in the practice of flow loop tuning. The results of the ISE optimisation are presented in Table 4.5.

One can see that the optimal tuning rules presented in Table 4.5 agree with the results obtained with the linear model of the flow process (Tables 3.5–3.8). However, there is a tendency to higher integral action at higher gain margins as per the rules based on the nonlinear model. The results of the simulation of the optimally tuned flow loop for different values of the gain margin are presented in Figs. 4.5, 4.6, 4.7 and 4.8. Response to the step change of the set point corresponding to 10 % of valve travel is applied at 30 s (it is assumed that 100 % of the flow range corresponds to 100 % of the valve travel). In Figs. 4.5 and 4.6, the step responses of the same system tuned in accordance with the Ziegler–Nichols closed-loop method and nonoptimal (sample) MRFT tuning rules of paper [19] with $\gamma_m = 3$ are also shown for comparison, respectively.

4.3 Conclusions

A flow control loop model that accounts for nonlinearities of the I/P transducer and pneumatic actuator is developed and presented. A nonconventional Lyapunov

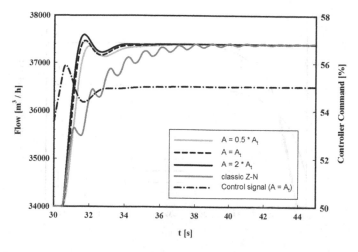

Fig. 4.5 Step response of flow loop with ISE-optimal PI controller; gain margin $\gamma_m = 2$

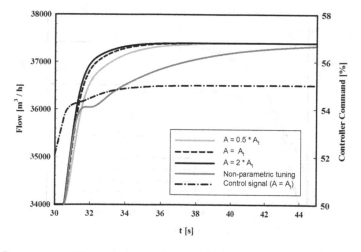

Fig. 4.6 Step response of flow loop with ISE-optimal PI controller; gain margin $\gamma_m = 3$

linearisation of the I/P transducer and actuator model is presented. It is shown that the Lyapunov linearisation results in reduction of model order and loss of some important properties of the model, which does not allow one to obtain operating point and relay amplitude-dependent tuning rules. It is demonstrated through the developed nonlinear model, the presented analysis and an illustrative example that the resulting tuning parameters would depend on the operating point and selection of the relay amplitude.

It can be seen from the presented example that the loop dynamics have a weak dependence on the operating point (valve opening in a steady state) and strong dependence on the relay amplitude. Both these effects are a result of the air flow between

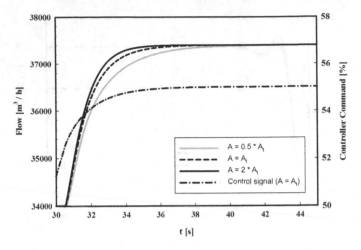

Fig. 4.7 Step response of flow loop with ISE-optimal PI controller; gain margin $\gamma_m = 4$

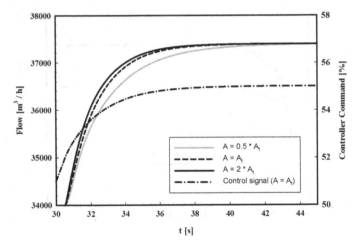

Fig. 4.8 Step response of flow loop with ISE-optimal PI controller; gain margin $\gamma_m = 5$

the I/P transducer and the valve. The former can be explained by the difference in volumes of the air chamber of the valve at different valve positions, so that if volume is larger, larger amounts of air intake are needed to achieve the same pressure increment. This in turn requires more time at the same air flow rate and results in slower dynamics. The latter effect is also related to the air flow dynamics. When larger amplitudes of the relay control are used and the I/P transducer provides larger pressure increments, air flow is still not proportional to the pressure differential, which results in larger values of *apparent time constants* of the air flow dynamics at a larger pressure differential between the I/P transducer and the valve. When this pressure differential becomes negligibly small the apparent time constant of the air flow dynamics approaches zero, and the third-order dynamics reduces to second-order dynamics.

As a result of the analysis carried out, it can be recommended that higher gain margins must be selected to ensure stability and satisfactory performance of the control loop in all operating conditions. For example, if the MRFT were carried out with 5 % relay amplitude but actual operating conditions required variations of the control within the range of ±20 % from the steady state, with respective fluctuations of the process variable, then the integral time constant of the PI controller might have to be increased by ~20 %. The increase of the integral time constant increases stability margins, and it is the time constant that has to be increased, as follows from Tables 4.3 and 4.4, but the proportional gain does not have to be decreased.

The solution of the optimisation problem with respect to c_1 and c_2 using the nonlinear model of the flow loop shows that the optimal values of these parameters differ from those in Chap. 3 using the linear model. The results obtained in this chapter using the nonlinear model are not comprehensive, of course. Rather, they demonstrate a point about the design of process-specific optimal tuning rules: The use of nonlinear models of the processes for non-parametric tuning may result in a number of efficient and useful solutions in the practice of loop tuning. However, practitioners may draw a useful methodology from the provided results involving some corrections to the tuning coefficients, which depend on the used amplitude of the MRFT/RFT and the operating point.

Chapter 5
Exact Model of MRFT and Parametric Tuning

Despite the primary subject of this book being non-parametric tuning, the modified relay feedback test presented earlier can also be used for parametric tuning. In this case the use of an exact model of the oscillation instead of one based on the approximate describing function method would be beneficial. This chapter aims to provide models suitable for parametric methods of tuning. Exact models of oscillations in the system with the RFT and the MRFT are presented in this chapter. The analysis is based on the locus of a perturbed relay system method. An exact model of oscillations in the system with the two-relay control is presented too. Orbital stability of the oscillations is analysed. An example illustrating identification of process dynamics through the provided models is given.

5.1 Locus of a Perturbed Relay System (LPRS) as Frequency-Domain Characteristic of Process

5.1.1 From Describing Function Analysis to LPRS Analysis

Identification of process model parameters from the relay feedback test has received much attention from researchers and found a number of applications. The idea proposed in [53] was later further developed in [21, 29, 34–36, 41, 45–47, 50, 58–60, 71, 78, 79, 83, 85, 87] and other publications (see also [44, 82]). Accuracy of identification on the one hand depends on the fidelity of the process model, and on the other hand on the accuracy of the model of the oscillations in the relay feedback system that is produced by inclusion of the relay in the loop. If both are precise enough then the conditions for accurate identification are propitious. In most publications, like the approach of the previous chapters of this book, the model of the oscillations is based on the describing function (DF) method. However, the DF method is approximate and a precise model of the oscillatory process in a relay feedback system would be an important component in the approach to identification of process parameters.

I. Boiko, *Non-parametric Tuning of PID Controllers*, Advances in Industrial Control, 97
DOI 10.1007/978-1-4471-4465-6_5, © Springer-Verlag London 2013

Considering the relay feedback principle on a larger scale one can see that relay feedback systems are one of the most important types of nonlinear systems. Applications of the relay feedback principle have evolved from vibrational voltage regulators and missile thruster servomechanisms of the 1940s to numerous on-off process parameter closed-loop control systems, sigma-delta modulators, process identification and automatic tuning of PID controller techniques, DC motors, hydraulic and pneumatic servo systems, and so on. It is enough to mention an enormous number of residential temperature control systems available throughout the world to understand how popular relay control systems are. A number of industrial examples of relay systems were given in the classic book on relay systems by Tsypkin [80]. The use of relay control provides a number of advantages over the use of linear control. Those advantages are: simplicity of design, cheaper components and ability to adapt the open-loop gain in the relay feedback system as parameters of the system change [8, 39, 55]. As a rule, they also provide a higher open-loop gain and a better performance [42] than do linear control systems. In some applications smoothing of the Coulomb friction and of other plant or process nonlinearities can also be achieved.

The theory of relay control systems has consistently received much attention from the worldwide research community. Traditionally, research problems include analysis of the existence and finding parameters of periodic motions, orbital stability and input-output problem (set point tracking and disturbance attenuation). The theory of relay feedback systems is presented in a number of classical and recent publications, some of which are given in the list of references.

The *locus of a perturbed relay system* (LPRS) method of analysis presented in this chapter is similar from the methodological point of view to the DF method and is designed to imitate the methodology of analysis used in the DF-based approach. Some concepts (like the notion of the *equivalent gain*) are also similar. However, the presented method is exact and the notions that are traditionally used within the DF method are redefined, so that in the exact approach they describe the system properties in the exact sense. Because of this obvious connection between the LPRS and the DF method, the former is presented as a logical extension of the latter. Some introductory material of the chapter is thus devoted to the DF method.

It is a well-known fact that relay feedback systems exhibit self-excited oscillations as their inherent mode of operation. Analysis of the frequency and amplitude of these oscillations is an objective of relay systems analysis. However, in the application of relay feedback systems, the *autonomous* mode, wherein no external signals are applied to the system, does not normally occur. An external signal always exists either in the form of an exogenous disturbance or a set point. In the first case, the system is supposed to respond to this disturbance in such a way as to provide its compensation. In the second case, the system is supposed to respond to the input so that the output can be brought in alignment with this external input. In both cases, the problem of analysis of the effect of external signals on system characteristics is an integral part of system performance analysis.

One can see that the problem of analysis of external signal propagation applies to both tracking and regulation. Moreover, if we consider a model of the system, these two types of systems would differ only by the point of external signal application, and from the perspective of the methodology of analysis they are the same. Since we are going to deal with models in this chapter we can consider only one signal applied to the system and consider it being a disturbance or a reference input signal depending on the system task. Naturally, the problem of propagation of external signals cannot be solved without the autonomous mode analysis being carried out first. Therefore, a complex analysis, which is supposed to include analysis of the autonomous mode and analysis of external signal propagation, needs to be carried out in practical applications.

In a general case (that includes time-delay linear process) the single-input-single-output (SISO) relay feedback system can be described by the following equations:

$$\begin{aligned} \dot{\mathbf{x}}(t) &= \mathbf{A}\mathbf{x}(t) + \mathbf{B}u(t - \tau), \\ y(t) &= \mathbf{C}\mathbf{x}(t), \end{aligned} \tag{5.1}$$

where $\mathbf{A} \in R^{n \times n}$, $\mathbf{B} \in R^{n \times 1}$ and $\mathbf{C} \in R^{1 \times n}$ are matrices, $\mathbf{A}$ is nonsingular, $\mathbf{x} \in R^{n \times 1}$ is the state vector, $y \in R^1$ is the system output, τ is the time delay or dead time (which can be set to zero if no time delay is present) and $u \in R^1$ is the control defined as follows:

$$u(t) = \begin{cases} +h & \text{if } e(t) = r_0 - y(t) \geq b \\ & \text{or } e(t) > -b, \ u(t-) = h \\ -h & \text{if } e(t) = r_0 - y(t) \leq -b \\ & \text{or } e(t) < b, \ u(t-) = -h \end{cases} \tag{5.2}$$

where r_0 is a constant input to the system, e is the error signal, h is the relay amplitude, $2b$ is the hysteresis value of the relay and $u(t-) = \lim_{\epsilon \to 0, \epsilon > 0} u(t - \epsilon)$ is the control at time instant immediately preceding time t. We shall consider that time $t = 0$ corresponds to the time of the error signal becoming equal to the positive half-hysteresis value (subject to $\dot{e} > 0$): $e(0) = b$ and call this time the *time of switch initiation*.

We represent the relay feedback system as a block diagram (Fig. 5.1). In the figure, $W_l(s)$ is the transfer function of the linear part of the relay system (the process if the system does not contain linear compensators or filters), which can be obtained from the matrix-vector description (5.1) as

$$W_l(s) = e^{-\tau s} \mathbf{C}(\mathbf{I}s - \mathbf{A})^{-1} \mathbf{B}. \tag{5.3}$$

In most situation the transfer function $W_l(s)$ is the transfer function of the process $W_p(s)$ (including a sensor, an actuator and a valve). However, $W_l(s)$ may also include some dynamics contained not in the process but in the controller, such as noise-reducing low-pass filters and additional lag-lead or lead-lag compensators. $W_l(s)$ thus contains all linear dynamics of the system (loop).

We shall assume that the linear part is strictly proper, i.e., the relative degree of $W_l(s)$ is 1 or higher, which is a valid assumption for all physically realisable systems.

Fig. 5.1 Relay feedback
system

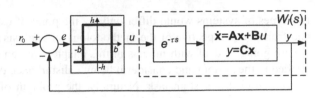

5.1.2 Symmetric Oscillations in Relay Feedback Systems

Relay feedback system may have a steady mode featuring non-vanishing self-excited oscillations, which is also often referred to as a self-excited periodic motion or a *limit cycle*. If the system does not have asymmetric nonlinearities this periodic motion in the autonomous mode (no external input applied) is symmetric. However, if a constant external input (disturbance) is applied to the system having a periodic motion the self-excited oscillations become biased or asymmetric. Because the control still can possess only two values h and $-h$, the asymmetry of the control signal is revealed as unequally-spaced switching, or pulses of different length for positive and negative parts.

The key approach to the analysis of relay feedback systems involves analysis of symmetric and asymmetric self-excited oscillations. The following sections give a general methodology of such an analysis based on frequency-domain concepts.

First we consider the autonomous mode of system operation ($r(t) \equiv 0$). Due to the character of the nonlinearity that results in the control having only two possible values, the system in Fig. 5.1 cannot have an equilibrium point. We can assume that a symmetric periodic process of unknown frequency Ω and amplitude a of the input to the relay occurs in the system. Finding the values of the frequency and amplitude is one of the main objectives of the analysis of relay feedback systems.

Very often, an approximate DF method is used in engineering practice. The main concepts of this method were developed in the 1940s and 1950s [55]; a comprehensive coverage of the DF method is given in [8, 39]. The DF method provides a simple and efficient but not precise solution of the periodic problem. For this method to provide an acceptable accuracy, the linear part of the system must be a low-pass filter that filters out higher harmonics, so that the input to the relay could approximately be considered a sinusoid.

In accordance with DF method concepts, we assume that the input to the nonlinearity is an harmonic signal and find the so-called describing function of the nonlinearity as a function of the amplitude and frequency using the following definition:

$$N(a,\omega) = \frac{\omega}{\pi a} \int_0^{2\pi/\omega} u(t) \sin \omega t \, dt + j \frac{\omega}{\pi a} \int_0^{2\pi/\omega} u(t) \cos \omega t \, dt. \qquad (5.4)$$

The DF given by formula (5.4) is in fact a complex gain that describes the transformation of the harmonic input by the nonlinearity with respect to the first harmonic in the output signal. For the hysteretic relay nonlinearity, the formula of the DF can be obtained analytically [8]:

$$N(a) = \frac{4h}{\pi a}\sqrt{1 - \left(\frac{b}{a}\right)^2} - j\frac{4hb}{\pi a^2}, \quad a \geq b. \tag{5.5}$$

For the hysteretic relay, the DF is a function of the amplitude only and does not depend on the frequency. The periodic solution in the relay feedback system can be found from the equation of the *harmonic balance* [8]:

$$W_l(j\Omega) = -\frac{1}{N(a)}, \tag{5.6}$$

which is a complex equation with two unknown values: frequency Ω and amplitude a. The expression on the left-hand side of equation (5.6) is the frequency response (i.e., Nyquist plot) of the linear part of the system. The value on the right-hand side is the negative reciprocal of the DF of the hysteretic relay. Let us obtain the respective expression from formula (5.5):

$$-N^{-1}(a) = -\frac{\pi a}{4h}\sqrt{1 - \left(\frac{b}{a}\right)^2} - j\frac{\pi b}{4h}, \quad a \geq b. \tag{5.7}$$

One can see from (5.7) that the imaginary part does not depend on the amplitude a, which results in a simple solution to (5.6). Graphically, the periodic solution of the equations of the relay feedback system would correspond to the point of intersection of the Nyquist plot of the linear part (being a function of the frequency) and of the negative reciprocal of the DF of the hysteretic relay (being a function of the amplitude) given by formula (5.7), in the complex plane. This periodic solution is not exact, which is a result of the approximate nature of the DF method itself that is based upon the assumption about the harmonic shape of the input signal to the relay. However, if the linear part of the system has the property of the low-pass filter, so that the higher harmonics of the control signal are attenuated well enough, the DF method may give a relatively precise result.

5.1.3 Asymmetric Oscillations in Relay Feedback Systems and Propagation of External Constant Inputs

Now we turn to the analysis of asymmetric oscillations in the relay feedback system, which is the key step to the analysis of the system response to constant and slow varying disturbances and set points.

Assume that the input to the system is a constant signal $r_0 : r(t) \equiv r_0$. Then an asymmetric periodic motion occurs in the system (Fig. 5.2), so that each signal now has a periodic term with zero mean value and a nonzero constant term: $u(t) = u_0 + u_p(t)$, $y(t) = y_0 + y_p(t)$, $e(t) = e_0 + e_p(t)$, where subscript 0 refers to the constant term, and subscript p refers to the periodic term of the function. If we apply

Fig. 5.2 Asymmetric
oscillations at unequally
spaced switches

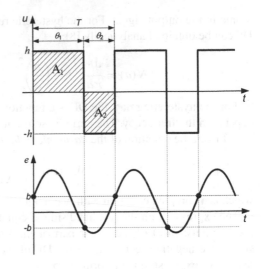

Fig. 5.3 Bias function and
equivalent gain

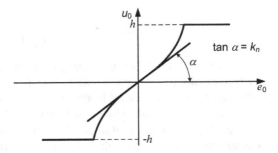

the Fourier series expansion then the periodic term will be the sum of all periodic
terms of the Fourier series. The constant term is the mean or averaged value of the
signal on the period.

We now quasi-statically (slowly) slew the constant input signal r_0 from a certain
negative value to a positive value so that at each value of the input signal the system
would exhibit a stable oscillation, and we measure the values of the constant term of
the control signal (mean control, which can be quantitatively represented in Fig. 5.2
as the area A_1 minus the area A_2) versus the constant term of the error signal (mean
error). Doing so we determine the constant term of the control signal as a function
of the constant term of the error signal, which would be not a discontinuous but
a smooth function: $u_0 = u_0(e_0)$. We shall further refer to this function as the *bias
function*. A typical bias function is depicted in Fig. 5.3. The described smoothing
effect is known as the *chatter smoothing* phenomenon [42]. The derivative of the
mean control with respect to the mean error taken in the point of zero mean error
$e_0 = 0$ (corresponding to zero constant input) provides the *equivalent gain* of the relay k_n, which conceptually is similar to the so-called *incremental gain* [8, 39] of the
DF method. The *equivalent gain* of the relay is thus used as a local approximation
of the *bias function*: $k_n = du_0/de_0|_{e_0=0} = \lim_{r_0 \to 0}(u_0/e_0)$.

Now we carry out analysis of asymmetric oscillations in the system of Fig. 5.1 caused by a non-zero constant input $r(t) \equiv r_0 \neq 0$. The DF of the hysteretic relay with a biased sine input is represented by the following well-known formula [8]:

$$N(a, e_0) = \frac{2h}{\pi a}\left[\sqrt{1 - \left(\frac{b + e_0}{a}\right)^2} + \sqrt{1 - \left(\frac{b - e_o}{a}\right)^2}\right] - j\frac{4hb}{\pi a^2}, \quad a \geq b + |e_0|,$$
(5.8)

where a is the amplitude of the oscillations. The mean control as a function of a and e_0 is given by the following formula:

$$u_0(a, e_0) = \frac{h}{\pi}\left(\arcsin\frac{b + e_0}{a} - \arcsin\frac{b - e_0}{a}\right).$$
(5.9)

From (5.8) and (5.9), we can obtain the DF of the relay and the derivative of the mean control with respect to the mean error for the symmetric sine input:

$$k_{n(DF)} = \frac{\partial u_0}{\partial e_0}\bigg|_{e_0=0} = \frac{2h}{\pi a}\frac{1}{\sqrt{1 - \left(\frac{b}{a}\right)^2}},$$
(5.10)

which is the *equivalent gain* of the relay (the subscript DF refers to the describing function method used for finding this characteristic).

Since at slow inputs, the relay feedback system behaves similarly to a linear system with respect to the response to those input signals, finding the *equivalent gain* value is the main point of input-output analysis. Once it is found, all subsequent analysis of the propagation of slow input signals can be carried out exactly as for a linear system. For that purpose the relay in the relay feedback system must be replaced with the *equivalent gain* determined above. The model obtained via the replacement of the relay with the *equivalent gain* would represent the model of the averaged (on the period of the oscillations) motions in the system.

5.2 Introduction to the LPRS

5.2.1 Computation of LPRS

As we just have seen, motions in relay feedback systems are normally analysed as motions in two separate dynamic subsystems: the "slow" subsystem and the "fast" subsystem. The "fast" subsystem describes self-excited oscillations or periodic motions. The "slow" subsystem deals with the forced motions caused by an input signal or by a disturbance and the component of the motion due to nonzero initial conditions. It usually describes the averaged (on the period of the self-excited oscillation) motion. The two dynamic subsystems interact with each other via a set of parameters: the results of the solution of the "fast" subsystem are used by the "slow" subsystem. This decomposition of dynamics is possible if the external input is much slower than the self-excited oscillations, which is normally the case. As within the

DF method, we shall proceed from the assumption that the external signals applied to the system are slow in comparison with the oscillations.

Consider again the harmonic balance equation (5.6). With the use of formulas for the negative reciprocal of the DF (5.7) and the equivalent gain of the relay (5.10), we can rewrite formula (5.6) as follows:

$$W_l(j\Omega) = -\frac{1}{2}\frac{1}{k_{n(DF)}} + j\frac{\pi}{4h}y_{(DF)}(0). \tag{5.11}$$

In the imaginary part of (5.11), we consider that the condition of the switch of the relay when it goes from minus to plus (defined as zero time) is the equality of the system output to the negative half-hysteresis: $y_{(DF)}(t = 0) = -b$. It follows from (5.8), (5.10) and (5.11) that the frequency of the oscillations and the equivalent gain in the system can be varied by changing the hysteresis value $2b$ of the relay. Therefore, the following two mappings can be considered: $M_1 : b \to \Omega$, $M_2 : b \to k_n$. Assume that M_1 has an inverse mapping (it follows from (5.8), (5.10) and (5.11) for the DF analysis and is proved below via deriving an analytical formula of that mapping) $M_1^{-1} : \Omega \to b$. Applying the chain rule, consider the mapping $M_2(M_1^{-1}) : \Omega \to b \to k_n$. Now let us define a certain function J exactly as the expression on the right-hand side of formula (5.11) but require from this function that the values of the equivalent gain and the output at zero time be exact values. Applying mapping $M_2(M_1^{-1}) : \omega \to b \to k_n$, $\omega \in [0, \infty)$, in which we treat frequency ω as an independent parameter, we write the following definition of this function:

$$J(\omega) = -\frac{1}{2}\frac{1}{k_n} + j\frac{\pi}{4h}y(t)|_{t=0}, \tag{5.12}$$

where $k_n = M_2(M_1^{-1}(\omega))$, $y(t)|_{t=0} = M_1^{-1}(\omega)$, $t = 0$ is the time of the switch of the relay from "$-h$" to "$+h$". Thus, $J(\omega)$ comprises the two mappings and is defined as a characteristic of the response of the linear part to the unequally spaced square-pulse input $u(t)$ subject to $r_0 \to 0$ as the frequency ω is varied. The real part of $J(\omega)$ contains information about the gain k_n, and the imaginary part of $J(\omega)$ comprises the condition of the switching of the relay and, consequently, contains information about the frequency of the oscillations. If we derive the function that satisfies the above requirements we will be able to obtain the exact values of the frequency of the oscillations and of the *equivalent gain*.

We will refer to the function $J(\omega)$ defined above and to its plot in the complex plane (with the frequency ω varied) as the *locus of a perturbed relay system* (LPRS). With LPRS of a given system computed, we are able to determine the frequency of the oscillations (as well as the amplitude) and the equivalent gain k_n (Fig. 5.4): The point of intersection of the LPRS and the line, which lies at distance $\pi b/(4h)$ below (if $b > 0$) or above (if $b < 0$) the horizontal axis and parallel to it (line "$-\pi b/4h$"), allows for computation of the frequency of the oscillations and of the equivalent gain k_n of the relay. According to (5.12), the frequency Ω of the oscillations can be computed via solving the equation:

$$\mathrm{Im}\, J(\Omega) = -\frac{\pi b}{4h} \tag{5.13}$$

Fig. 5.4 LPRS and analysis
of relay feedback system

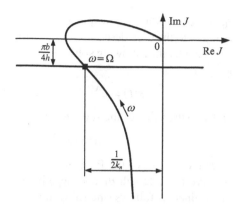

(i.e., $y(0) = -b$ is the condition of the relay switch) and the gain k_n can be computed as:

$$k_n = -\frac{1}{2\,\mathrm{Re}\,J(\Omega)}. \tag{5.14}$$

Formula (5.13) provides a periodic solution and is, therefore, a *necessary condition of the existence of a periodic motion* in the system.[1] Formula (5.12) is only a definition and not intended for the purpose of computing the LPRS $J(\omega)$. It is shown below that although $J(\omega)$ is defined via the parameters of the oscillations in a closed-loop system, it can be easily derived from the parameters of the linear part without employing the variables of formula (5.12).

5.2.2 Computation of the LPRS from Differential Equations

5.2.2.1 Matrix State Space Description Approach

We now derive the LPRS formula for a time-delay process in accordance with its definition (5.12) from the system description given by formulas (5.1) and (5.2). Derivation of the LPRS formula necessitates finding a periodic solution of (5.1) and (5.2), which for an arbitrary nonlinearity is a fundamental problem in both mathematics and control theory. Its solution can be traced back to the works of Poincaré, who introduced a widely used geometric interpretation of this problem as finding a closed orbit in the state space. The approach of Poincaré involves finding a certain map that relates subsequent intersection by the state trajectory of a certain hyperplane. A fixed point of this map would give a periodic solution in the dynamic system. We can define the hyperplane for our analysis by the following equation:

---

[1]The actual existence of a periodic motion depends on a number of other factors too, including orbital stability of the obtained periodic solution and initial conditions.

$r_0 - \mathbf{C}\mathbf{x} = b$, which corresponds to the initiation of the relay switch from $-h$ to $+h$ (subject to the time derivative of the error signal being positive).

Consider the following solution for the state vector subject to the constant control $u = \pm 1$:

$$\mathbf{x}(t) = e^{\mathbf{A}(t-\tau)}\mathbf{x}(\tau) \pm \mathbf{A}^{-1}(e^{\mathbf{A}(t-\tau)} - \mathbf{I})\mathbf{B}, \quad t > \tau$$

and also the value of the state vector at time $t = \tau$:

$$\mathbf{x}(\tau) = e^{\mathbf{A}\tau}\mathbf{x}(0) - \mathbf{A}^{-1}(e^{\mathbf{A}\tau} - \mathbf{I})\mathbf{B}.$$

A fixed point of the Poincaré return map for asymmetric periodic motion with positive pulse length θ_1 and negative pulse length θ_2 and the unity amplitude is determined as follows (the LPRS does not depend on the control amplitude, which can be assumed as $h = 1$):

$$\boldsymbol{\eta}_p = e^{\mathbf{A}\theta_1}\boldsymbol{\rho}_p + \mathbf{A}^{-1}(e^{\mathbf{A}\theta_1} - \mathbf{I})\mathbf{B}, \tag{5.15}$$

$$\boldsymbol{\rho}_p = e^{\mathbf{A}\theta_2}\boldsymbol{\eta}_p - \mathbf{A}^{-1}(e^{\mathbf{A}\theta_2} - \mathbf{I})\mathbf{B}, \tag{5.16}$$

where $\boldsymbol{\rho}_p = \mathbf{x}(\tau) = \mathbf{x}(T + \tau)$, $\boldsymbol{\eta}_p = \mathbf{x}(\theta_1 + \tau)$ ($\boldsymbol{\rho}_p \in R^n$, $\boldsymbol{\eta}_p \in R^n$) and T is the period of the oscillations. We suppose that θ_1 and θ_2 are known. Then (5.15) and (5.16) can be solved for $\boldsymbol{\rho}_p$ and $\boldsymbol{\eta}_p$ as follows. Substitute (5.16) for $\boldsymbol{\rho}_p$ in (5.15):

$$\boldsymbol{\eta}_p = e^{\mathbf{A}\theta_1}\left[e^{\mathbf{A}\theta_2}\boldsymbol{\eta}_p - \mathbf{A}^{-1}(e^{\mathbf{A}\theta_2} - \mathbf{I})\mathbf{B}\right] + \mathbf{A}^{-1}(e^{\mathbf{A}\theta_1} - \mathbf{I})\mathbf{B},$$

which leads to

$$\boldsymbol{\eta}_p = e^{\mathbf{A}(\theta_1+\theta_2)}\boldsymbol{\eta}_p - e^{\mathbf{A}\theta_1}\mathbf{A}^{-1}(e^{\mathbf{A}\theta_2} - \mathbf{I})\mathbf{B} + \mathbf{A}^{-1}(e^{\mathbf{A}\theta_1} - \mathbf{I})\mathbf{B}.$$

Regroup the above equation collecting the terms containing $\boldsymbol{\eta}_p$ on the left-hand side:

$$(\mathbf{I} - e^{\mathbf{A}(\theta_1+\theta_2)})\boldsymbol{\eta}_p = \left[-e^{\mathbf{A}\theta_1}\mathbf{A}^{-1}(e^{\mathbf{A}\theta_2} - \mathbf{I}) + \mathbf{A}^{-1}(e^{\mathbf{A}\theta_1} - \mathbf{I})\right]\mathbf{B}.$$

Express $\boldsymbol{\eta}_p$ from the above formula:

$$\boldsymbol{\eta}_p = (\mathbf{I} - e^{\mathbf{A}(\theta_1+\theta_2)})^{-1}\left[-e^{\mathbf{A}\theta_1}\mathbf{A}^{-1}(e^{\mathbf{A}\theta_2} - \mathbf{I}) + \mathbf{A}^{-1}(e^{\mathbf{A}\theta_1} - \mathbf{I})\right]\mathbf{B}.$$

Considering that $\mathbf{A}e^{\mathbf{A}t}\mathbf{A}^{-1} = e^{\mathbf{A}t}$ when simplifying the above expression we obtain the following formula:

$$\boldsymbol{\eta}_p = (\mathbf{I} - e^{\mathbf{A}(\theta_1+\theta_2)})^{-1}\mathbf{A}^{-1}\left[2e^{\mathbf{A}\theta_1} - e^{\mathbf{A}(\theta_1+\theta_2)} - \mathbf{I}\right]\mathbf{B}.$$

Similarly substitute (5.15) into (5.16) to derive an expression for $\boldsymbol{\rho}_p$:

$$\boldsymbol{\rho}_p = e^{\mathbf{A}(\theta_1+\theta_2)}\boldsymbol{\rho}_p + e^{\mathbf{A}\theta_2}\mathbf{A}^{-1}(e^{\mathbf{A}\theta_1} - \mathbf{I})\mathbf{B} - \mathbf{A}^{-1}(e^{\mathbf{A}\theta_2} - \mathbf{I})\mathbf{B}$$

or

$$(\mathbf{I} - e^{\mathbf{A}(\theta_1+\theta_2)})\boldsymbol{\rho}_p = e^{\mathbf{A}\theta_2}\mathbf{A}^{-1}(e^{\mathbf{A}\theta_1} - \mathbf{I})\mathbf{B} - \mathbf{A}^{-1}(e^{\mathbf{A}\theta_2} - \mathbf{I})\mathbf{B},$$

which gives

$$\boldsymbol{\rho}_p = (\mathbf{I} - e^{\mathbf{A}(\theta_1+\theta_2)})^{-1}\mathbf{A}^{-1}\left[e^{\mathbf{A}(\theta_1+\theta_2)} - 2e^{\mathbf{A}\theta_2} + \mathbf{I}\right]\mathbf{B}.$$

Considering that $\theta_1 + \theta_2 = T$, solution of (5.15), (5.16) results in:

$$\boldsymbol{\eta}_p = (\mathbf{I} - e^{\mathbf{A}T})^{-1}\mathbf{A}^{-1}[2e^{\mathbf{A}\theta_1} - e^{\mathbf{A}T} - \mathbf{I}]\mathbf{B}, \tag{5.17}$$

$$\boldsymbol{\rho}_p = (\mathbf{I} - e^{\mathbf{A}T})^{-1}\mathbf{A}^{-1}[e^{\mathbf{A}T} - 2e^{\mathbf{A}\theta_2} + \mathbf{I}]\mathbf{B}. \tag{5.18}$$

Now, considering that

$$\boldsymbol{\rho}_p = \mathbf{x}(\tau) = e^{\mathbf{A}\tau}\mathbf{x}(0) - \mathbf{A}^{-1}(e^{\mathbf{A}\tau} - \mathbf{I})\mathbf{B}$$

and

$$\boldsymbol{\eta}_p = \mathbf{x}(\theta_1 + \tau) = e^{\mathbf{A}\tau}\mathbf{x}(\theta_1) + \mathbf{A}^{-1}(e^{\mathbf{A}\tau} - \mathbf{I})\mathbf{B},$$

we find $\mathbf{x}(0)$ as follows:

$$\mathbf{x}(0) = e^{-\mathbf{A}\tau}\left[(\mathbf{I} - e^{\mathbf{A}T})^{-1}\mathbf{A}^{-1}[e^{\mathbf{A}T} - 2e^{\mathbf{A}\theta_2} + \mathbf{I}]\mathbf{B} + \mathbf{A}^{-1}(e^{\mathbf{A}\tau} - \mathbf{I})\mathbf{B}\right]. \tag{5.19}$$

Reasoning along similar lines, find the formula for $\mathbf{x}(\theta_1)$.

$$\mathbf{x}(\theta_1) = e^{-\mathbf{A}\tau}\left[(\mathbf{I} - e^{\mathbf{A}T})^{-1}\mathbf{A}^{-1}[2e^{\mathbf{A}\theta_1} - e^{\mathbf{A}T} - \mathbf{I}]\mathbf{B} - \mathbf{A}^{-1}(e^{\mathbf{A}\tau} - \mathbf{I})\mathbf{B}\right]. \tag{5.20}$$

Consider now the symmetric motion as a limit of (5.19) at $\theta_1, \theta_2 \to \theta = T/2$:

$$\lim_{\theta_1,\theta_2 \to \theta = T/2} \mathbf{x}(0) = (\mathbf{I} + e^{\mathbf{A}T/2})^{-1}\mathbf{A}^{-1}[\mathbf{I} + e^{\mathbf{A}T/2} - 2e^{\mathbf{A}(T/2-\tau)}]\mathbf{B}. \tag{5.21}$$

The imaginary part of the LPRS can be obtained from (5.21) in accordance with its definition as follows:

$$\begin{aligned}
\operatorname{Im} J(\omega) &= \frac{\pi}{4}\mathbf{C}\lim_{\theta_1,\theta_2 \to \theta = T/2} \mathbf{x}(0) \\
&= \frac{\pi}{4}\mathbf{C}(\mathbf{I} + e^{\mathbf{A}\pi/\omega})^{-1}(\mathbf{I} + e^{\mathbf{A}\pi/\omega} - 2e^{\mathbf{A}(\pi/\omega-\tau)})\mathbf{A}^{-1}\mathbf{B}. \tag{5.22}
\end{aligned}$$

For deriving the expression of the real part of the LPRS consider the periodic solution (5.17) and (5.18) as a result of the feedback action

$$\begin{cases} r_0 - y(0) = b, \\ r_0 - y(\theta_1) = -b. \end{cases} \tag{5.23}$$

Having solved the set of equations (5.23) for r_0 we can obtain:

$$r_0 = \frac{y(0) + y(\theta_1)}{2}.$$

Hence, the constant term of the error signal $e(t)$ is

$$e_0 = r_0 - y_0 = \frac{y(0) + y(\theta_1)}{2} - y_0.$$

The real part of the LPRS definition formula can be transformed into:

$$\operatorname{Re} J(\omega) = -\frac{1}{2}\lim_{\gamma \to \frac{1}{2}} \frac{0.5[y(0) + y(\theta_1)] - y_0}{u_0}, \tag{5.24}$$

where $\gamma = \frac{\theta_1}{\theta_1+\theta_2} = \frac{\theta_1}{T}$. Then $\theta_1 = \gamma T$, $\theta_2 = (1-\gamma)T$, $u_0 = 2\gamma - 1$, and (5.24) can be rewritten as:

$$\mathrm{Re}\, J(\omega) = -\frac{1}{2} \lim_{\gamma \to \frac{1}{2}} \frac{0.5C[\mathbf{x}(0) + \mathbf{x}(\theta_1)] - y_0}{2\gamma - 1}, \qquad (5.25)$$

where $\mathbf{x}(0)$ and $\mathbf{x}(\theta_1)$ are given by (5.19) and (5.20), respectively. Now, first considering the limit

$$\lim_{\gamma \to \frac{1}{2}} \frac{e^{\mathbf{A}\gamma T} - e^{-\mathbf{A}\gamma T} e^{\mathbf{A}T}}{2\gamma - 1} = \mathbf{A} T e^{\mathbf{A}T/2} \qquad (5.26)$$

find the following limit.

$$\lim_{u_0 \to 0 \, (\theta_1+\theta_2=T=const)} \frac{\mathbf{x}(0) + \mathbf{x}(\theta_1)}{u_0} = \lim_{\gamma \to \frac{1}{2}} \frac{\mathbf{x}(0) + \mathbf{x}(\theta_1)}{2\gamma - 1}$$

$$= 2e^{-\mathbf{A}\tau} T \left(\mathbf{I} - e^{\mathbf{A}T}\right)^{-1} e^{\mathbf{A}T/2} \mathbf{B}. \quad (5.27)$$

To find $\lim_{u_0 \to 0} \frac{y_0}{u_0}$ consider the equations for the constant terms of the variables (i.e., averaged variables), which are obtained from the original equations of the process via equating the derivatives to zero:

$$\begin{cases} 0 = \mathbf{A}\mathbf{x}_0 + \mathbf{B}u_0, \\ y_0 = \mathbf{C}\mathbf{x}_0. \end{cases}$$

From these equations obtain $\mathbf{x}_0 = -\mathbf{A}^{-1}\mathbf{B}u_0$ and $y_0 = -\mathbf{C}\mathbf{A}^{-1}\mathbf{B}u_0$. Therefore,

$$\lim_{u_0 \to 0} \frac{y_0}{u_0} = -\mathbf{C}\mathbf{A}^{-1}\mathbf{B}, \qquad (5.28)$$

which is the gain of the process. The real part of the LPRS is obtained by substituting (5.27) and (5.28) for respective limits in (5.25).

$$\mathrm{Re}\, J(\omega) = -\frac{1}{2} \lim_{u_0 \to 0} \frac{\frac{1}{2}\mathbf{C}[\mathbf{x}(0) + \mathbf{x}(\theta_1)] - y_0}{u_0}$$

$$= -\frac{1}{2}T\mathbf{C}\left(\mathbf{I} - e^{\mathbf{A}T}\right)^{-1} e^{\mathbf{A}(T/2-\tau)}\mathbf{B} - \frac{1}{2}\mathbf{C}\mathbf{A}^{-1}\mathbf{B}. \qquad (5.29)$$

Finally, the state space description based formula of the LPRS can be obtained by combining formulas (5.21) and (5.29) as follows:

$$J(\omega) = -\frac{1}{2}\mathbf{C}\left[\mathbf{A}^{-1} + \frac{2\pi}{\omega}\left(\mathbf{I} - e^{\frac{2\pi}{\omega}\mathbf{A}}\right)^{-1} e^{\left(\frac{\pi}{\omega}-\tau\right)\mathbf{A}}\right]\mathbf{B}$$

$$+ j\frac{\pi}{4}\mathbf{C}\left(\mathbf{I} + e^{\frac{\pi}{\omega}\mathbf{A}}\right)^{-1}\left(\mathbf{I} + e^{\frac{\pi}{\omega}\mathbf{A}} - 2e^{\left(\frac{\pi}{\omega}-\tau\right)\mathbf{A}}\right)\mathbf{A}^{-1}\mathbf{B}. \qquad (5.30)$$

The real part of the LPRS was derived under the assumption of the limits at $u_0 \to 0$ and at $\gamma \to \frac{1}{2}$ being equal. This can only be true if the derivative dT/du_0 at the point $u_0 = 0$ is zero, which follows from the symmetry of the function $T = T(u_0)$. A rigorous proof of this is given in [15].

5.2.2.2 Orbital Asymptotic Stability

The stability of periodic orbits (limit cycles) is usually referred to as the *orbital stability*. The notion of orbital stability is different from the notion of the stability of an equilibrium point because in an orbitally stable motion the difference between the perturbed and unperturbed motions does not necessarily vanish. What is important is that, if orbitally stable, the motion in the perturbed system converges to the orbit of the unperturbed system. In the practice of application of the RFT/MRFT to parametric tuning, analysis of orbital stability of the oscillations using the process model serves the feasibility study because only the oscillations that are orbitally stable can be realised in the test. And if the use of a certain process model results in orbitally unstable oscillations in RFT or MRFT then either the process model is incorrect or the test is unsuitable for the particular process.

There are two different approaches to analysing orbital stability. The first one involves analysis of the distance between the trajectory point representing the current value of the state vector and the set representing the trajectory of the periodic motion. The second approach is based on the design and analysis of Poincaré maps. In relay feedback systems, analysis of orbital stability can be conveniently reduced to the analysis of certain equivalent discrete-time system with time instants corresponding to the switches of the relay, which can be obtained from the original system by considering the Poincaré map of the motion having an initial perturbation. If we assume that the initial state is $\mathbf{x}(0) = \rho + \delta\rho$, where $\delta\rho$ is the initial perturbation, and find the mapping $\delta\rho \to \delta\eta$, we will be able to make a conclusion about the orbital stability of the system from consideration of the Jacobian matrix of this mapping. The local orbital stability criterion can be formulated as the following statement.

Local orbital stability. The solution of system (5.1), (5.2) provided by condition (5.13) is locally orbitally asymptotically stable if all eigenvalues of matrix

$$\Phi_0 = \left[\mathbf{I} - \frac{\mathbf{v}(\tau + \frac{T}{2}-)\mathbf{C}}{\mathbf{C}\mathbf{v}(\tau + \frac{T}{2}-)}\right]e^{\mathbf{A}\frac{T}{2}}, \tag{5.31}$$

where the velocity vector is given by

$$\mathbf{v}\left(\tau + \frac{T}{2}-\right) = \dot{\mathbf{x}}\left(\tau + \frac{T}{2}-\right) = 2\left(\mathbf{I} + e^{\mathbf{A}T/2}\right)^{-1}e^{\mathbf{A}(T/2-\tau)}\mathbf{B},$$

have magnitudes less than *one*.

Proof Consider the process with time $t \in (-\infty, \infty)$ and assume that by the time $t = 0$, where time $t = 0$ is the switch initiation time, a periodic motion has been established, with the fixed point of the Poincaré map given by formulas (5.17) and (5.18). Also, assume that there are no switches of the relay within the interval $(0, \tau)$. At the switch time $t = \tau$ the state vector receives a perturbation (deviation from the value in a periodic motion): $\mathbf{x}(\tau) = \rho = \rho_p + \delta\rho$. Find the mapping of this perturbation into the perturbation of the state vector at the time of the next switch and analyse whether the initial perturbation vanishes. For the time interval

$t \in [\tau, t^*]$, where t^* is the time of the switch of the relay from $+1$ to -1, the state vector (while the control is $u = 1$) is given as follows:

$$\mathbf{x}(t) = e^{\mathbf{A}(t-\tau)}(\rho_p + \delta\rho) + \mathbf{A}^{-1}\left(e^{\mathbf{A}(t-\tau)} - \mathbf{I}\right)\mathbf{B}$$
$$= e^{\mathbf{A}(t-\tau)}\rho_p + \mathbf{A}^{-1}\left(e^{\mathbf{A}(t-\tau)} - \mathbf{I}\right)\mathbf{B} + e^{\mathbf{A}(t-\tau)}\delta\rho, \qquad (5.32)$$

where the first two addends give the unperturbed motion, and the third addend gives the motion due to the initial perturbation $\delta\rho$. Let us denote

$$\mathbf{x}(t^*) = \eta = \eta_p + \delta\eta. \qquad (5.33)$$

It should be noted that $\boldsymbol{\eta}_p$ is determined not at time $t = t^*$ but at time $t = \tau + \theta_1$. The main task of our analysis is to find the mapping $\delta\rho \rightarrow \delta\eta$. Therefore, it follows from (5.33) that

$$\delta\eta = \mathbf{x}(t^*) - \boldsymbol{\eta}_p. \qquad (5.34)$$

Considering that all perturbations are small and times t^* and $\tau + \theta_1$ are close, evaluate $\mathbf{x}(t^*)$ via linear approximation of $\mathbf{x}(t)$ at the point $t = \tau + \theta_1$.

$$\mathbf{x}(t^*) - \mathbf{x}(\tau + \theta_1) = \dot{\mathbf{x}}(\tau + \theta_1 -)(t^* - \tau - \theta_1),$$

where $\dot{\mathbf{x}}(\tau + \theta_1 -)$ is the value of the derivative at time immediately preceding time $t = \tau + \theta_1$. Express $\mathbf{x}(t^*)$ from the last equation as follows.

$$\mathbf{x}(t^*) = \mathbf{x}(\tau + \theta_1) + \dot{\mathbf{x}}(\tau + \theta_1 -)(t^* - \tau - \theta_1). \qquad (5.35)$$

Now evaluate $\mathbf{x}(\tau + \theta_1)$ using (5.32).

$$\mathbf{x}(\tau + \theta_1) = \eta_p + e^{\mathbf{A}\theta_1}\delta\rho. \qquad (5.36)$$

Substitute (5.35) and (5.36) into (5.34).

$$\delta\eta = \mathbf{x}(\tau + \theta_1) + \dot{\mathbf{x}}(\tau + \theta_1 -)(t^* - \tau - \theta_1) - \eta_p$$
$$= \mathbf{v}(\tau + \theta_1 -)(t^* - \tau - \theta_1) + e^{\mathbf{A}\theta_1}\delta\rho, \qquad (5.37)$$

where $\mathbf{v}(t) = \dot{\mathbf{x}}(t)$. Now find $(t^* - \tau - \theta_1)$ from the switch condition.

$$(t^* - \tau - \theta_1)\dot{y}(\tau + \theta_1 -) = -\delta y(\tau + \theta_1) = -\mathbf{C}\delta\mathbf{x}(\tau + \theta_1 -) = -\mathbf{C}e^{\mathbf{A}\theta_1}\delta\rho. \qquad (5.38)$$

From (5.38), we now obtain

$$t^* - \tau - \theta_1 = -\frac{\delta y(\tau + \theta_1)}{\dot{y}(\tau + \theta_1 -)} = -\frac{\mathbf{C}e^{\mathbf{A}\theta_1}}{\dot{y}(\tau + \theta_1 -)}\delta\rho.$$

Substitute the expression for $t^* - \tau - \theta_1$ into (5.37).

$$\delta\eta = -\mathbf{v}(\tau + \theta_1 -)\frac{\mathbf{C}e^{\mathbf{A}\theta_1}}{\dot{y}(\tau + \theta_1 -)}\delta\rho + e^{\mathbf{A}\theta_1}\delta\rho$$
$$= \left[\mathbf{I} - \frac{\mathbf{v}(\tau + \theta_1 -)\mathbf{C}}{\mathbf{C}\mathbf{v}(\tau + \theta_1 -)}\right]e^{\mathbf{A}\theta_1}\delta\rho. \qquad (5.39)$$

Denote the Jacobian matrix of the mapping $\delta\rho \rightarrow \delta\eta$ as follows:

$$\Phi_1 = \left[\mathbf{I} - \frac{\mathbf{v}(\tau + \theta_1-)\mathbf{C}}{\mathbf{Cv}(\tau + \theta_1-)}\right]e^{\mathbf{A}\theta_1}. \tag{5.40}$$

In order to assess local orbital stability of the periodic solution, we need to find the Jacobian matrix of the Poincaré return map, which will be the chain rule application of two mappings: $\delta\rho \rightarrow \delta\eta$ and $\delta\eta \rightarrow \delta\rho$. However, for the symmetric motion, checking only a half period of the motion is sufficient, which gives formula (5.31). $\square$

In addition to the stability analysis, it can be recommended that the direction of the relay switch should be verified, too. This condition is formulated as the satisfaction of the following inequality:

$$\dot{y}\left(\frac{T}{2}-\right) = \mathbf{Cv}\left(\frac{T}{2}-\right) > 0,$$

where $\mathbf{v}\left(\frac{T}{2}-\right)$ is given above.

5.2.3 Computation of the LPRS from Process Transfer Function

5.2.3.1 Infinite Series Approach

A different formula for $J(\omega)$ can now be derived for the case of the linear part given by a transfer function. Suppose that the linear part does not have integrators and write the Fourier series expansion of the signal $u(t)$ (Fig. 5.2):

$$u(t) = u_0 + 4h/\pi \sum_{k=1}^{\infty} \sin\big(\pi k\theta_1/(\theta_1 + \theta_2)\big)/k$$
$$\times \big\{\cos(k\omega\theta_1/2)\cos(k\omega t) + \sin(k\omega\theta_1/2)\sin(k\omega t)\big\},$$

where $u_0 = h(\theta_1 - \theta_2)/(\theta_1 + \theta_2)$, $\omega = 2\pi/(\theta_1 + \theta_2)$. Propagation of each harmonic in $u(t)$ through the linear part features multiplication by a factor equal to the magnitude frequency response at frequency $k\omega$ and a phase shift equal to the phase frequency response at frequency $k\omega$. Therefore, $y(t)$ as a response of the linear part with the transfer function $W_l(s)$ can be written as:

$$y(t) = y_0 + 4h/\pi \sum_{k=1}^{\infty} \sin\big(\pi k\theta_1/(\theta_1 + \theta_2)\big)/k$$
$$\times \big\{\cos(k\omega\theta_1/2)\cos\big[k\omega t + \varphi_l(k\omega)\big]$$
$$+ \sin(k\omega\theta_1/2)\sin\big[k\omega t + \varphi_l(k\omega)\big]\big\}A_l(k\omega), \tag{5.41}$$

where $\varphi_l(k\omega) = \arg W_l(jk\omega)$, $A_l(k\omega) = |W_l(jk\omega)|$, $y_0 = u_0|W_l(j0)|$. The conditions of the switches of the relay have the form of equations (5.23) where $y(0)$ and $y(\theta_1)$ can be obtained from (5.41) if we set $t = 0$ and $t = \theta_1$ respectively:

$$y(0) = y_0 + 4h/\pi \sum_{k=1}^{\infty} [0.5 \sin(2\pi k\theta_1/(\theta_1 + \theta_2))\mathrm{Re}W_l(jk\omega)$$

$$+ \sin^2(\pi k\theta_1/(\theta_1 + \theta_2))\mathrm{Im}W_l(jk\omega)]/k, \tag{5.42}$$

$$y(\theta_1) = y_0 + 4h/\pi \sum_{k=1}^{\infty} [0.5 \sin(2\pi k\theta_1/(\theta_1 + \theta_2))\mathrm{Re}W_l(jk\omega)$$

$$- \sin^2(\pi k\theta_1/(\theta_1 + \theta_2))\mathrm{Im}W_l(jk\omega)]/k. \tag{5.43}$$

Differentiating (5.23) with respect to r_0 (and taking into account (5.42) and (5.43)) we obtain the formulas containing the derivatives at the point $\theta_1 = \theta_2 = \theta = \pi/\omega$. Having solved those equations for $\mathrm{d}(\theta_1 - \theta_2)/\mathrm{d}r_0$ and $\mathrm{d}(\theta_1 + \theta_2)/\mathrm{d}r_0$ we shall obtain: $\mathrm{d}(\theta_1 + \theta_2)/\mathrm{d}r_0|_{r_0=0} = 0$, which corresponds to the derivative of the frequency of the oscillations and the derivative of the pulse length difference:

$$\frac{\mathrm{d}(\theta_1 - \theta_2)}{\mathrm{d}r_0}\bigg|_{r_0=0} = \frac{2\theta}{h(|W_l(0)| + 2\sum_{k=1}^{\infty}\cos(\pi k)\mathrm{Re}W_l(\omega k))}. \tag{5.44}$$

Considering that the derivative of the pulse length difference is related to the equivalent gain and applying the formula of the closed-loop system transfer function we can write:

$$\frac{\mathrm{d}(\theta_1 - \theta_2)}{\mathrm{d}r_0}\bigg|_{r_0=0} = \frac{k_n}{1 + k_n|A_l(0)|}\frac{2\theta}{h}. \tag{5.45}$$

Having solved together equations (5.44) and (5.45) for k_n we obtain the following expression:

$$k_n = \frac{0.5}{\sum_{k=1}^{\infty}(-1)^k \mathrm{Re}\, W_l(k\pi/\theta)}. \tag{5.46}$$

Taking into account formula (5.46), the identity $\omega = \pi/\theta$ and the definition of the LPRS (5.12) we obtain the final form of expression for $\mathrm{Re}\, J(\omega)$. Similarly, having solved the set of equations (5.23) where $\theta_1 = \theta_2 = \theta$, and $y(0)$ and $y(\theta_1)$ are given by (5.42) and (5.43), respectively, we obtain the final formula of $\mathrm{Im}\, J(\omega)$. Having put the real and imaginary parts together, we obtain the final formula of the LPRS $J(\omega)$ for relay feedback systems:

$$J(\omega) = \sum_{k=1}^{\infty}(-1)^{k+1} \mathrm{Re}\, W_l(k\omega) + j \sum_{k=1}^{\infty}\frac{1}{2k - 1} \mathrm{Im}\, W_l[(2k - 1)\omega]. \tag{5.47}$$

5.2.3.2 Partial Fraction Expansion Technique

Another technique of LPRS computation is based on the possibility of derivation of analytical formulas for the LPRS of low-order dynamics. It follows from (5.47) that the LPRS possesses the property of *additivity*, which can be formulated as follows.

Additivity property. If the transfer function $W_l(s)$ of the linear part is a sum of n transfer functions: $W_l(s) = W_1(s) + W_2(s) + \cdots + W_n(s)$ then the LPRS $J(\omega)$ can be calculated as a sum of the n component LPRS: $J(\omega) = J_1(\omega) + J_2(\omega) + \cdots + J_n(\omega)$, where $J_i(\omega)$ $(i = 1, \ldots, n)$ is the LPRS of the relay system with the transfer function of the linear part being $W_i(s)$.

The considered property offers a technique of the LPRS computation based on the expansion of the process transfer function into partial fractions. Indeed, if $W_l(s)$ is expanded into the sum of first and second order dynamics then LPRS $J(\omega)$ can be calculated via summation of the component LPRS $J_i(\omega)$ corresponding to each of the addends in the transfer function expansion, subject to available analytical formulas for the LPRS of first and second order dynamics. Formulas for $J(\omega)$ of first and second order dynamics are presented in Table 5.1. MATLAB code for the computation of the LPRS of low order dynamics is presented in Chap. 8. Loci of a perturbed relay system of some of these dynamics are considered below.

5.2.4 LPRS of Low Order Dynamics

5.2.4.1 LPRS of First Order Dynamics

It was mentioned above that one possible technique of LPRS computing is to represent the transfer function as partial fractions, compute the LPRS of the component transfer functions (the partial fractions) and add those component LPRS in accordance with the additivity property. To realise this methodology we have to know the formulas of the LPRS for first and second order dynamics. It is of similar meaning and importance as knowing the characteristics of first and second order dynamics in linear system analysis. The knowledge of the LPRS of the low order dynamics is important for other reasons, too, because some features of the LPRS of low order dynamics can be extended to higher order systems. Those features are considered in the following section.

The LPRS formula for the first order dynamics given by the transfer function $W(s) = K/(Ts + 1)$ is given by the following formula [15]:

$$J(\omega) = \frac{K}{2}\left(1 - \frac{\pi}{T\omega}\operatorname{csch}\frac{\pi}{T\omega}\right) - j\frac{\pi K}{4}\tanh\frac{\pi}{2\omega T}, \qquad (5.48)$$

where csch(.) and tanh(.) are hyperbolic cosecant and tangent, respectively.

The plot of the LPRS for $K = 1$, $T = 1$ is given in Fig. 5.5. The point $(0.5K, -j\frac{\pi}{4}K)$ corresponds to the frequency $\omega = 0$ and the point $(0, j0)$ corresponds to the frequency $\omega = \infty$. The high frequency segment of the LPRS has the imaginary axis for its asymptote.

With the formula for the LPRS in hand, we can easily find the frequency of periodic motions in the relay feedback system, with the linear part being the first order dynamics. The LPRS is a continuous function of the frequency, and for every hysteresis value from the range $b \in [0, hK]$ there exists a periodic solution of the frequency that can be determined from (5.13) and (5.48), which is

Table 5.1 Formulas of the LPRS $J(\omega)$

Transfer fun. $W(s)$	LPRS $J(\omega)$ and MATLAB function
$\frac{K}{s}$	$0 - j\frac{\pi^2 K}{8\omega}$ MATLAB function: **lprsint(k,w)**
$\frac{K}{Ts+1}$	$\frac{K}{2}(1 - \alpha\,\text{csch}\,\alpha) - j\frac{\pi K}{4}\tanh(\alpha/2)$ $\alpha = \pi/(T\omega)$ MATLAB function: **lprs1ord(k,t,w)**
$\frac{Ke^{-\tau s}}{Ts+1}$	$\frac{K}{2}(1 - \alpha e^\gamma\,\text{csch}\,\alpha) + j\frac{\pi K}{4}(\frac{2e^{-\alpha}e^\gamma}{1+e^{-\alpha}} - 1)$ $\alpha = \frac{\pi}{T\omega},\ \gamma = \frac{\tau}{T}$ MATLAB function: **lprsfopdt(k,t,tau,w)**
$\frac{K}{(T_1 s+1)(T_2 s+1)}$	$\frac{K}{2}[1 - T_1/(T_1 - T_2)\alpha_1\,\text{csch}\,\alpha_1 - T_2/(T_2 - T_1)\alpha_2\,\text{csch}\,\alpha_2)]$ $-j\frac{\pi K}{4}/(T_1 - T_2)[T_1\tanh(\alpha_1/2) - T_2\tanh(\alpha_2/2)]$ $\alpha_1 = \pi/(T_1\omega),\ \alpha_2 = \pi/(T_2\omega)$
$\frac{K}{s^2+2\xi s+1}$	$\frac{K}{2}[(1 - (B + \gamma C)/(\sin^2\beta + \sinh^2\alpha)]$ $-j\frac{\pi K}{4}(\sinh\alpha - \gamma\sin\beta)/(\cosh\alpha + \cos\beta)$ $\alpha = \pi\xi/\omega,\ \beta = \pi(1 - \xi^2)^{1/2}/\omega,\ \gamma = \alpha/\beta$ $B = \alpha\cos\beta\sinh\alpha + \beta\sin\beta\cosh\alpha,$ $C = \alpha\sin\beta\cosh\alpha - \beta\cos\beta\sinh\alpha$ MATLAB function: **lprs2ord1(k,xi,w)**
$\frac{Ks}{s^2+2\xi s+1}$	$\frac{K}{2}[\xi(B + \gamma C) - \pi/\omega\cos\beta\sinh\alpha]/(\sin^2\beta + \sinh^2\alpha)]$ $-j\frac{\pi K}{4}(1 - \xi^2)^{-1/2}\sin\beta/(\cosh\alpha + \cos\beta)$ $\alpha = \pi\xi/\omega,\ \beta = \pi(1 - \xi^2)^{1/2}/\omega,\ \gamma = \alpha/\beta$ $B = \alpha\cos\beta\sinh\alpha + \beta\sin\beta\cosh\alpha,$ $C = \alpha\sin\beta\cosh\alpha - \beta\cos\beta\sinh\alpha$ MATLAB function: **lprs2ord2(k,xi,w)**
$\frac{Ks}{(s+1)^2}$	$\frac{K}{2}[\alpha(-\sinh\alpha + \alpha\cosh\alpha)/\sinh^2\alpha - j0.25\pi\alpha/(1 + \cosh\alpha)]$ $\alpha = \pi/\omega$ MATLAB function: **lprs2ord3(k,w)**
$\frac{Ks}{(T_1 s+1)(T_2 s+1)}$	$\frac{K}{2}/(T_2 - T_1)[\alpha_2\,\text{csch}\,\alpha_2 - \alpha_1\,\text{csch}\,\alpha_1]$ $-j\frac{\pi K}{4}/(T_2 - T_1)[\tanh(\alpha_1/2) - \tanh(\alpha_2/2)]$ $\alpha_1 = \pi/(T_1\omega),\ \alpha_2 = \pi/(T_2\omega)$ MATLAB function: **lprs2ord4(k,xi,w)**

$\Omega = \frac{\pi}{2T}\text{arctanh}(\frac{b}{hK})$. It is easy to show that when the hysteresis value b tends to zero then the frequency of the periodic solution tends to infinity ($\lim_{b\to 0}\Omega = \infty$) and when the hysteresis value b tends to hK then the frequency of the periodic solution tends to zero ($\lim_{b\to cK}\Omega = 0$). From (5.48), we can also see that the imaginary

Fig. 5.5 LPRS of first order
dynamics

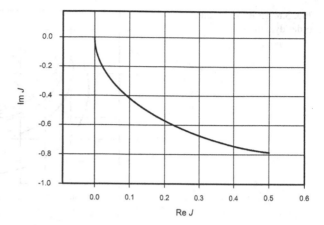

part of the LPRS is a monotone function of the frequency. Therefore, the condition
of the existence of a finite frequency periodic solution holds for any nonzero hysteresis value from the specified range, and the limit for $b \to 0$ exists and corresponds
to infinite frequency.

It is easy to show that the oscillations are always orbitally stable. The stability
of a periodic solution can be verified via finding eigenvalues of the Jacobian of
the corresponding Poincaré map. For the first order system the only eigenvalue of
this Jacobian will always be zero, as there is only one system variable, which also
determines the condition of the switch of the relay.

5.2.4.2 LPRS of Second Order Dynamics

Similar analysis can now be carried out for the second order dynamics. Let the
matrix $\mathbf{A}$ of (5.1) be $\mathbf{A} = [0\ 1;\ -a_1\ -a_2]$ and the delay be $\tau = 0$. Here, consider a
few cases, all with $a_1 > 0, a_2 > 0$.

(i) Let $a_2^2 - 4a_1 < 0$. Then the process transfer function can be written as $W(s) = K/(T^2 s^2 + 2\xi T s + 1)$. The LPRS formula can be found, for example, via
   expanding the above transfer function into partial fractions and applying to it
   formula (5.48) obtained for the first order dynamics. However, the coefficients
   of those partial fractions will be complex numbers and this must be considered.
   The formula of the LPRS for the second-order dynamics can be written as
   follows:

$$J(\omega) = \frac{K}{2}\left(1 - \frac{g + \gamma w}{\sin^2 \beta + \sinh^2 \alpha}\right) - j\frac{\pi K}{4}\frac{\sinh \alpha - \gamma \sin \beta}{\cosh \alpha + \cos \beta}, \qquad (5.49)$$

where $\alpha = \frac{\pi \xi}{\omega T}$, $\beta = \frac{\pi \sqrt{1-\xi^2}}{\omega T}$, $\gamma = \alpha/\beta$, $g = \alpha \cos \beta \sinh \alpha + \beta \sin \beta \cosh \alpha$
and $w = \alpha \sin \beta \cosh \alpha + \beta \cos \beta \sinh \alpha$. The plots of the LPRS for $K = 1$,
$T = 1$ and different values of damping ξ are given in Fig. 5.6 (Curve 1, $\xi = 1$,
Curve 2, $\xi = 0.85$, Curve 3, $\xi = 0.7$, Curve 4, $\xi = 0.55$ and Curve 5, $\xi = 0.4$).

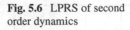

Fig. 5.6 LPRS of second
order dynamics

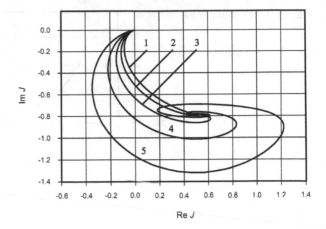

Re *J*

The high frequency segment of the LPRS of the second order process approaches the real axis.

The following limits are easily found: $\lim_{\omega \to \infty} J(\omega) = (0, j0)$ and $\lim_{\omega \to 0} J(\omega) = (0.5K, -j\frac{\pi}{4}K)$. These limits give the two boundary points of the LPRS corresponding to zero frequency and infinite frequency. Analysis of function (5.49) shows that it does not have intersections with the real axis except at the origin. Since $J(\omega)$ is a continuous function of the frequency ω (following from formula (5.49)) a solution of equation (5.13) exists for any $b \in (0, hK)$. Therefore, a periodic solution of finite frequency exists for the considered second order system for every value of b within the specified range, and there is a periodic solution of infinite frequency for $b = 0$.

(ii) Consider the case when $a_2^2 - 4a_1 = 0$. To obtain the LPRS formula, we can use formula (5.49) and find the limit for $\xi \to 1$. The LPRS for this case is given in Fig. 5.6 (Curve 1). All subsequent analysis and conclusions are the same as in case (i).

(iii) Assume that $a_2^2 - 4a_1 > 0$. The transfer function can be expanded into two partial fractions, and then according to the *additivity property* the LPRS can be computed as a sum of the two components. The subsequent analysis is similar to the previous one.

(iv) For $a_1 = 0$ the transfer function is $W(s) = K/[s(Ts + 1)]$. For this process the LPRS is given by the following formula, which can be obtained via partial fraction expansion technique:

$$J(\omega) = \frac{K}{2}\left(\frac{\pi}{T\omega}\operatorname{csch}\frac{\pi}{T\omega} - 1\right) + j\frac{\pi K}{4}\left(\tanh\frac{\pi}{2\omega T} + \frac{\pi}{2\omega}\right). \qquad (5.50)$$

The plot of the LPRS for $K = 1$, $T = 1$ is given in Fig. 5.7. The entire plot is located in the 3rd quadrant. The point $(0.5K, -j\infty)$ corresponds to the frequency $\omega = 0$ and the point $(0, j0)$ corresponds to the frequency $\omega = \infty$. The high frequency segment of the LPRS has the real axis for an asymptote.

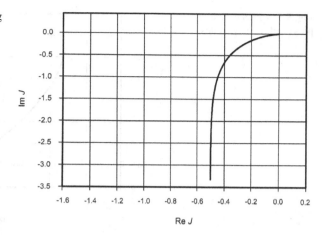

Fig. 5.7 LPRS of integrating second order dynamics

5.2.4.3 LPRS of First Order Plus Dead Time Dynamics

One more model is of particular interest because many industrial processes can be adequately approximated by the first order plus time delay (dead time) transfer function $W(s) = \frac{Ke^{-\tau s}}{Ts+1}$, where K is a gain, T is a time constant and τ is a time delay (dead time). This factor results in the importance of analysis of this model. The formula of the LPRS for the first order plus dead time transfer function can be derived as follows [15]:

$$J(\omega) = \frac{K}{2}\left(1 - \alpha e^{\gamma}\operatorname{csch}\alpha\right) + j\frac{\pi}{4}K\left(\frac{2e^{-\alpha}e^{\gamma}}{1+e^{-\alpha}} - 1\right) \qquad (5.51)$$

where $\gamma = \tau/T$. The plots of the LPRS for $\gamma = 0$ (Curve 1), $\gamma = 0.2$ (Curve 2), $\gamma = 0.5$ (Curve 3), $\gamma = 1.0$ (Curve 4) and $\gamma = 1.5$ (Curve 5) are depicted in Fig. 5.8. All the plots begin at the point $(0.5, -j\pi/4)$ corresponding to the frequency $\omega = 0$. Plot 1 (corresponding to zero dead time) arrives at the origin, which corresponds to infinite frequency. Other plots are defined only for the frequencies that are less than the frequency corresponding to half of the period.

5.2.5 Some Properties of the LPRS

We now consider some properties of the LPRS that may be helpful for LPRS computing, checking the results of computing and especially for the design of linear compensators. One of the most important properties is the additivity property. A few other properties that relate to the boundary points corresponding to zero frequency and infinite frequency are considered below. We now find the coordinates of the initial point of the LPRS (which corresponds to zero frequency) considering formula (5.30). For that purpose, we find the limit of function $J(\omega)$ for ω tending to zero. First we evaluate two limits: $\lim_{\omega \to 0}[\frac{2\pi}{\omega}(\mathbf{I} - e^{\frac{2\pi}{\omega}\mathbf{A}})^{-1}e^{\frac{\pi}{\omega}\mathbf{A}}] = \mathbf{0}$,

Fig. 5.8 LPRS of first-order
plus dead time dynamics

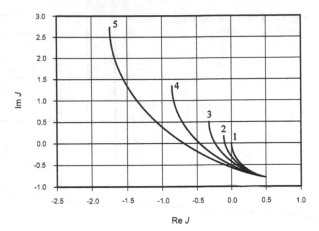

$\lim_{\omega \to 0}[(\mathbf{I} + e^{\frac{\pi}{\omega}\mathbf{A}})^{-1}(\mathbf{I} - e^{\frac{\pi}{\omega}\mathbf{A}})] = \mathbf{I}$. Knowing these, we can write the limit for the LPRS as follows:

$$\lim_{\omega \to 0} J(\omega) = \left[-0.5 + j\frac{\pi}{4}\right]\mathbf{CA}^{-1}\mathbf{B}. \tag{5.52}$$

The product of matrices $\mathbf{CA}^{-1}\mathbf{B}$ in (5.52) is the negative value of the gain of the process transfer function. Therefore, the following statement has been proven. For a *nonintegrating linear part of the relay feedback system, the initial point of the corresponding LPRS is* $(0.5K, -j\pi/4K)$, *where K is the static gain of the linear part*. This completely agrees with the above analysis of the LPRS of the first and second order dynamics; see, for example, Fig. 5.5 and Fig. 5.6.

To find the limit of $J(\omega)$ for ω tending to infinity, consider the following two limits of the expansion into power series of the exponential function:

$$\lim_{\omega \to \infty} \exp\left(\frac{\pi}{\omega}\mathbf{A}\right) = \lim_{\omega \to \infty} \sum_{n=0}^{\infty} \frac{(\pi/\omega)^n}{n!}\mathbf{A}^n = \mathbf{I}$$

and

$$\lim_{\omega \to \infty}\left\{\frac{2\pi}{\omega}\left[\mathbf{I} - \exp\left(\frac{2\pi}{\omega}\mathbf{A}\right)\right]^{-1}\right\} = \lim_{\lambda = \frac{2\pi}{\omega} \to 0}\left\{\lambda\left[\mathbf{I} - \exp(\lambda\mathbf{A})\right]^{-1}\right\} = -\mathbf{A}^{-1}.$$

Taking these limits into account we prove that the endpoint of the LPRS for the nonintegrating linear part without time delay is the origin:

$$\lim_{\omega \to \infty, \tau = 0} J(\omega) = 0 + j0. \tag{5.53}$$

For time delay processes, the endpoint of the LPRS is also the origin, which can be shown using the infinite series formula (5.47). However, the approach of the LPRS to the origin at $\omega \to \infty$ is not asymptotic.

5.3 LPRS Model of Oscillations in MRFT

As before, assume that the process is given as follows:

$$\dot{x} = Ax + Bu,$$
$$y = Cx, \tag{5.54}$$

where $A \in R^{n \times n}$, $B \in R^{n \times 1}$ and $C \in R^{1 \times n}$ are matrices, $x \in R^n$, $y \in R^1$. And, the modified RFT algorithm is given as follows:

$$u(t) = \begin{cases} h & \text{if } e(t) \geq b_1 \text{ or } (e(t) > -b_2 \text{ and } u(t-) = h), \\ -h & \text{if } e(t) \leq -b_2 \text{ or } (e(t) < b_1 \text{ and } u(t-) = -h), \end{cases} \tag{5.55}$$

where $b_1 = -\beta e_{min}$, $b_2 = \beta e_{max}$, $e_{max} > 0$ and $e_{min} < 0$ are last singular points of the error signal corresponding to the last maximum and minimum values of $e(t)$ after crossing the zero level, $u(t-) = \lim_{\epsilon \to 0, \epsilon > 0} u(t - \epsilon)$ is the control value at the time immediately preceding current time t, h is the amplitude of the relay and β is a constant.

As before let us assume that the steady state behaviour of the system (5.54), (5.55) is a periodic, unimodal symmetric limit cycle with zero mean. It was shown above through the use of the DF method that this motion could exist and the MRFT algorithm for the case of a periodic motion would correspond to the hysteretic relay characteristic of the controller.

5.3.1 Exact Frequency-Domain Analysis of Oscillations

The DF analysis of the system with the MRFT given in Chap. 4 provides a simple and systematic, but approximate evaluation of the magnitude and frequency of the periodic motions in the closed-loop system (5.54)–(5.55) driven by the modified RFT algorithm. For parametric tuning, there is an opportunity for more precise identification of an underlying process model than the one based on the describing function model of oscillations. Consider now derivation of an exact model of the oscillations and input-output dependence in a system with the MRFT.

As noted above, despite the fact that in a periodic motion the MRFT acts as the relay system, the hysteresis value of this relay depends on the amplitude of the oscillation and is, therefore, unknown. Therefore, the conventional application of the LPRS method is impossible. Yet, an exact solution can be obtained via application of the following frequency-domain approach.

We design the complex function $\Phi(\omega)$ in a way that its use would imitate the use of the describing function technique in finding a possible periodic motion in a system with the MRFT. We introduce the following function, the magnitude of which at each point gives the amplitude of the oscillation in the system with the MRFT, at the corresponding frequency of oscillations:

$$\Phi(\omega) = -\sqrt{[a_y(\omega)]^2 - y^2\left(\frac{\pi}{\omega}, \omega\right)} + jy\left(\frac{\pi}{\omega}, \omega\right), \tag{5.56}$$

where $y(\frac{\pi}{\omega}, \omega)$ is the value of the system periodic output signal at the time instant when the relay switches from $-h$ to h (π/ω is a half period in the periodic motion and $t = 0$ is assumed, without loss of generality, to be the time of the relay switch from h to $-h$), and $a_y(\omega)$ is the amplitude of the process output signal in the assumed periodic motion of frequency ω:

$$a_y = \max_{t \in [0,T]} |y(t, \omega)|. \tag{5.57}$$

The periodic output signal $y(t, \omega)$ can be computed, for example, by means of its Fourier series:

$$
\begin{aligned}
y(t, \omega) &= \frac{4h}{\pi} \sum_{k=1}^{\infty} \frac{1}{k} \sin\left(\frac{1}{2}\pi k\right) \sin[k\omega t + \varphi(k\omega)] A_l(\omega k) \\
&= \frac{4h}{\pi} \sum_{k=1}^{\infty} \frac{(-1)^{k+1}}{2k - 1} \sin[(2k - 1)\omega t + \varphi_l((2k - 1)\omega)] \cdot A_l((2k - 1)\omega),
\end{aligned}
$$
$$\tag{5.58}$$

where $\varphi_l(\omega) = \arg W_l(j\omega)$, $A_l(\omega) = |W_l(j\omega)|$ are the phase and magnitude of $W_l(j\omega)$ at the frequency ω, respectively.

The frequency-dependent variable $a_y(\omega)$ can be computed by using (5.57) and (5.58), and $y(\frac{\pi}{\omega}, \omega)$ as the imaginary part of the LPRS (with coefficient $\frac{4h}{\pi}$) or via using the Fourier series (5.58). As a result, function $\Phi(\omega)$ has the same imaginary part as the imaginary part of the LPRS with the coefficient $4h/\pi$, and the magnitude of function $\Phi(\omega)$ at the intersection point represents the amplitude of the oscillation in a steady periodic motion.

Having computed the function $\Phi(\omega)$, we can carry out analysis of possible periodic motions the same way as we did above via the DF technique, simply replacing the Nyquist plot of $W(j\omega)$ with the function $\Phi(\omega)$ given by (5.56).

The methodology of the exact frequency-domain analysis is the same as within the DF analysis. Again, the point of intersection of the straight line drawn through the origin and making the angle with the horizontal axis equal to $\arcsin \beta$, as depicted in Fig. 5.9, and of the locus $\Phi(\omega)$ gives the frequency and the amplitude of the periodic motion.

5.3.2 Describing Function Analysis of External Signal Propagation

The autonomous properties of the MRFT algorithm were investigated above. It is shown superior in some respects to conventional relay control. However, pure autonomous modes never occur in real systems due to the existence of external disturbances. Therefore, analysis of transfer properties of systems with MRFT algorithm is important in terms of the robustness of the identification when the system is subject to external disturbances. A similar effect of asymmetric oscillations in the test

Fig. 5.9 Plot of function $\Phi(\omega)$ and finding a periodic solution

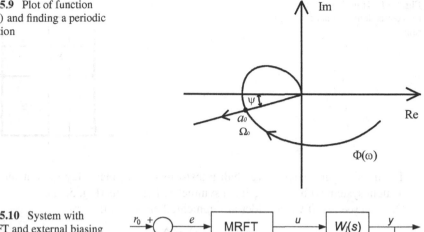

Fig. 5.10 System with MRFT and external biasing signal

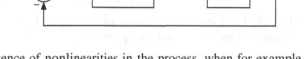

may occur due to the presence of nonlinearities in the process, when for example valve motion would be different in the opening and closing directions. These situations motivate us to build a model, which would describe the transfer properties of the systems with the MRFT algorithm.

Let us apply an approach similar to the one that was used for analysis of the transfer properties of the relay systems. First apply a constant input to the closed-loop system controlled by the MRFT algorithm; after that, considering this constant input an infinitesimally small value, determine the equivalent gain of the MRFT algorithm, and finally extend the obtained results to the case of variable inputs (slow in comparison with the oscillations). We reiterate here the results previously obtained for the so-called suboptimal second-order sliding mode control algorithm [23].

Let the system with the MRFT algorithm be given as the block diagram (Fig. 5.10).

An external constant input r_0 is applied to the system. With this arrangement, let us consider that the switching happens in accordance with the following equation:

$$u = h \operatorname{sign}(e + \beta e_{Mi}), \tag{5.59}$$

where e_{Mi} is the latest singular point of e, i.e., the value of e at the most recent time instant t_{Mi} ($i = 1, 2, \ldots$) such that $\dot{e}(t_{Mi}) = 0$, so that the switching instants depend on the amplitude of the error signal (not the system output). Obviously, in the autonomous mode the singular points of $e(t)$ coincide with those of $y(t)$, and the amplitudes of e and y are equal. Therefore, the value of the amplitude a_y determined above can be used instead of the value of the amplitude a_e of $e(t)$. Yet, at the input-output analysis with respect to the external constant input, the fact of the asymmetric amplitudes for positive and negative values of $e(t)$ has to be considered. Because of the hysteresis value being dependent on the amplitudes of the oscillation the switching points of the relay (values $b_1 > 0$ and $b_2 > 0$) become asymmetric, and we have to consider the relay with asymmetric hysteresis.

Fig. 5.11 Transformation
into equivalent symmetric
relay

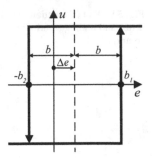

To simplify our analysis, we shall transform the original relay system into an
equivalent system with relay having a symmetric hysteresis (Fig. 5.11).

One can see that if we consider an augmented error $e^*(t)$:

$$e^*(t) = e(t) - \Delta_e, \tag{5.60}$$

where Δ_e is the shift of the vertical axis in Fig. 5.11 (the distance between the solid
vertical axis and the dashed one) given as

$$\Delta_e = \frac{b_1 - b_2}{2}, \tag{5.61}$$

then analysis of the relay system with symmetrical hysteretic relay having

$$b = \frac{b_1 + b_2}{2} \tag{5.62}$$

and error signal $e^*(t)$ can be carried out using the DF method (subject to accounting
for dependence of b on the amplitude of the oscillation). Therefore, we can reduce
the task of analysis to a symmetrical one by applying transformations (5.60)–(5.62)
to the system. Therefore, the system can be considered as one having a symmetric
hysteresis b and an additional input Δ_e due to error augmentation (5.60).

Now we find the dependence of Δ_e on the value of the constant input r_0. Consider
the Fourier expansion of the asymmetric periodic control $u(t)$:

$$u(t) = u_0 + \frac{4h}{\pi} \sum_{k=1}^{\infty} \frac{1}{k} \sin\left(\pi k \frac{\theta_1}{\theta_1 + \theta_2}\right)$$

$$\times \left\{\cos(k\omega\theta_1/2)\cos(k\omega t) + \sin(k\omega\theta_1/2)\sin(k\omega t)\right\},$$

where $u_0 = h(\theta_1 - \theta_2)/(\theta_1 + \theta_2)$, $\omega = 2\pi/(\theta_1 + \theta_2)$. Therefore, considering only
the fundamental frequency component (as per the filtering hypothesis) we can write
for (5.60):

$$u(t) \approx u_0 + \frac{4h}{\pi} \sin\left(\pi \frac{\theta_1}{\theta_1 + \theta_2}\right) \left\{\cos(\omega\theta_1/2)\cos(\omega t) + \sin(\omega\theta_1/2)\sin(\omega t)\right\},$$
$$\tag{5.63}$$

which can be rewritten as follows:

$$u(t) \approx u_0 + \frac{4h}{\pi} \sin\left(\frac{\pi\theta_1}{\theta_1 + \theta_2}\right) \cos(\omega(t - \theta_1/2)). \tag{5.64}$$

One can see from (5.64) that the amplitude of the oscillations of the control signal is

$$a_u = \frac{4h}{\pi} \sin\left(\frac{\pi\theta_1}{\theta_1 + \theta_2}\right)$$

(the fundamental frequency component). Therefore, application of the external constant signal r_0 results not only in the bias of the error signal $e(t)$ but also in the decrease of the amplitude of the oscillation $e(t)$. Moreover, b_1 and b_2 are different and depend on the positive and negative amplitude of $e(t)$, respectively. Considering the following relationship between the hysteresis and the amplitudes of the error signal oscillation:

$$b_1 = -\beta e_{\min},$$
$$b_2 = \beta e_{\max},$$

or

$$b_1 = -\beta r_0 + \beta a_p,$$
$$b_2 = \beta r_0 - \beta a_n,$$

where a_p is the "positive" amplitude of $y(t)$, $a_p > 0$ and a_n is the "negative" amplitude of $y(t)$, $a_n < 0$.

Therefore, $\Delta_e = -\frac{\beta}{2}(2r_0 - a_p - a_n)$ and the derivative $\frac{d\Delta_e}{du_0}$ is

$$\frac{d\Delta_e}{du_0} = \frac{\beta}{2}\left(\frac{da_p}{du_0} + \frac{da_n}{du_0} - 2\frac{dr_0}{du_0}\right). \tag{5.65}$$

The derivatives of (5.65) in the point corresponding to $u_0 = 0$ are going to be found below. It follows from (5.64) that

$$a_p = u_0 + \frac{4h}{\pi} \sin\left(\pi \frac{\theta_1}{\theta_1 + \theta_2}\right),$$

$$a_n = u_0 - \frac{4h}{\pi} \sin\left(\pi \frac{\theta_1}{\theta_1 + \theta_2}\right).$$

Also, considering the relationship between the averaged control u_0 and the positive pulse duration θ_1, which is $u_0 = h(\theta_1 - \theta_2)/(\theta_1 + \theta_2) = c(2\theta_1 - T)/T$, and therefore resulting in the derivative $\frac{du_0}{d\theta_1} = 2h/T$, we can obtain formulas for the following derivatives at the point $\theta_1 = \theta_2$ ($u_0 = 0$):

$$\frac{da_p}{du_0} = \frac{da_n}{du_0} = W_l(j0).$$

The derivative $\frac{dr_0}{du_0}$ in formula (5.65) can be obtained from the equation of the constant terms balance in the system:

$$(r_0 - u_0 W_l(j0))k_{nDF} = u_0,$$

where k_{nDF} is the equivalent gain of the MRFT algorithm, which we aim to find. The dependence of r_0 on u_0 is fictitious. The actual dependence is rather the opposite—of u_0 on r_0. The former dependence of r_0 on u_0 can only be interpreted

Fig. 5.12 Equivalent system
for averaged component
propagation through system
with MRFT

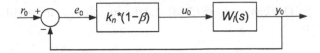

as revealing the answer to "how much should we adjust r_0 to obtain the change we need in u_0?". The subscript DF indicates that this is a variable derived with the use of the describing function method.

Therefore,

$$\frac{dr_0}{du_0} = \frac{1}{k_{nDF}} + W_l(j0)$$

and

$$\frac{d\Delta_e}{du_0} = \beta W_l(j0) - \beta\left(\frac{1}{k_{nDF}} + W_l(j0)\right) = -\frac{\beta}{k_{nDF}}. \tag{5.66}$$

Once the additional input due to the error augmentation is determined (formula (5.66)) we can obtain an analytical formula of the equivalent gain of the MRFT algorithm. Let

$$k_{nDF}^* = \frac{2h}{\pi\sqrt{a_{yDF}^2 - b^2}} = \frac{2h}{\pi a_{yDF}\sqrt{1 - \beta^2}}$$

be the equivalent gain of the hysteretic relay. It does not account for the Δ_e and is, therefore, not an equivalent gain of the whole algorithm. The equivalent gain of the algorithm can be determined as the equivalent gain of the relay k_{nDF}^* having a feedback with the gain $\frac{-\beta}{k_{nDF}}$:

$$k_{nDF} = \frac{k_{nDF}^*}{1 + \beta k_{nDF}^*/k_{nDF}}.$$

We can find from the last equation that

$$k_{nDF} = k_{nDF}^*(1 - \beta). \tag{5.67}$$

The rest of the analysis is the same as of the conventional relay system. The effect of the error augmentation via the gain (5.67) can be depicted as in Fig. 5.12.

The results obtained for the constant input r_0 can be extended to the case of slowly varying r_0, where "slow" implies that the frequency of the input signal is much lower than the frequency of oscillations in the system with the MRFT. In Fig. 5.12, the results obtained for the constant input r_0 are applied to slow varying input $r(t)$. As a result, the process gain $W_l(j0)$ is replaced with the process gain at the frequency of the input signal. One can see that application of the delayed switching as per the MRFT algorithm results in the decrease of the equivalent gain of the relay. This decrease depends on the parameter β of the MRFT.

5.3.3 *Exact Frequency-Domain Analysis of External Signal Propagation*

We carry out the same input-output analysis as above but via the LPRS method, considering the system with MRFT algorithm, to which an external constant input r_0 is applied as in Fig. 5.10.

Because of the hysteresis value being dependent on the amplitudes of the oscillation the switching points of the relay (values b_1 and b_2) become asymmetric, and we have to consider the relay with asymmetric hysteresis. To simplify our analysis by utilising the LPRS method we transform the original relay system into an equivalent system with relay having a symmetric hysteresis (Fig. 5.11) through adding one more input signal. We note that if we consider an augmented error $e^*(t)$, with Δ_e being the shift of the vertical axis in Fig. 5.11 (the distance between the solid vertical axis and the dashed line) then analysis of the relay system with the symmetrical hysteretic relay can be done through the LPRS method (subject to the account for the dependence of b on the amplitude of the oscillation). Therefore, we can simplify our analysis considering it symmetrical by applying transformations (5.60)–(5.62) to the system.

We now consider the dependence of the average error e_0 and the average control u_0 on the external constant input signal r_0 applied to the system. The simulations show that these two dependences are close to linear ones in a wide range. The dependence of u_0 on e_0 (due to the variation of r_0) will be referred to as the *bias function*. As a result of the above two functions being virtually linear, the *bias function* is also linear. Therefore, we can consider analysis of the propagation of the variables averaged on the period of the oscillation as analysis of a linear system through the approximation of the bias function by the equivalent gain of the relay. The equivalent gain is, therefore, the slope of the bias function at $r_0 = 0$. Now we can derive an expression for the *equivalent gain*.

We consider the unequally spaced control $u(t)$ of amplitude h with the positive and negative pulse durations being respectively θ_1 and θ_2 (the same way we did earlier in this chapter), and analyse the response of the linear part to this control. After that we can obtain the value of constant input r_0 and the relay hysteresis that is needed to produce the given $u(t)$ in the closed-loop system. To find the equivalent gain we find the reciprocal derivative:

$$\frac{de_0}{du_0} = \frac{1}{k_n}. \tag{5.68}$$

It follows from (5.60) that

$$\frac{de_0}{du_0} = \frac{de_0^*}{du_0} + \frac{d\Delta_e}{du_0} = \frac{1}{k_{nLPRS}} + \frac{d\Delta_e}{du_0}, \tag{5.69}$$

where k_{nLPRS} is the equivalent gain obtained via the LPRS for the equivalent system with symmetric hysteresis $2b$ (see the material of this chapter, presented above).

The second component in formula (5.69) is given by the expression (5.65). The last derivative in (5.65) can be written as follows (the same way as in the DF analysis):

$$\frac{dr_0}{du_0} = \frac{1}{k_n} + W_l(j0). \tag{5.70}$$

The first two derivatives in (5.65) are found below. The output of the process is given as follows:

$$y(t) = y_0 + \frac{4h}{\pi} \sum_{k=1}^{\infty} \sin\big(\pi k\theta_1/(\theta_1 + \theta_2)\big)/k$$
$$\times \Big\{\cos(k\omega\theta_1/2)\cos\big[k\omega t + \varphi_l(k\omega)\big]$$
$$+ \sin(k\omega\theta_1/2)\sin\big[k\omega t + \varphi_l(k\omega)\big]\Big\}A_l(k\omega). \tag{5.71}$$

Considering that $\theta_1 + \theta_2 = T = \frac{2\pi}{\omega}$ we can rewrite the last formula as follows:

$$y(t) = y_0 + \frac{4h}{\pi} \sum_{k=1}^{\infty} \frac{1}{k} \sin\left(\frac{k}{2}\omega\theta_1\right)$$
$$\times \Big\{\cos\left(\frac{k}{2}\omega\theta_1\right)\cos\big[k\omega t + \varphi_l(k\omega)\big]$$
$$+ \sin\left(\frac{k}{2}\omega\theta_1\right)\sin\big[k\omega t + \varphi_l(k\omega)\big]\Big\}A_l(k\omega). \tag{5.72}$$

We now find the derivative of $y(t)$ with respect to θ_1:

$$\frac{\partial y(t)}{\partial \theta_1} = \frac{\partial y_0}{\partial \theta_1} + \frac{4h}{\pi} \sum_{k=1}^{\infty} \frac{1}{k} \Big\{\frac{k}{2}\omega \cos\left(\frac{k}{2}\omega\theta_1\right)\Big\{\cos\left(\frac{k}{2}\omega\theta_1\right)\cos\big[k\omega t + \varphi_l(k\omega)\big]$$
$$+ \sin\left(\frac{k}{2}\omega\theta_1\right)\sin\big[k\omega t + \varphi_l(k\omega)\big]\Big\}A_l(k\omega)$$
$$+ \sin\left(\frac{k}{2}\omega\theta_1\right)\frac{k}{2}\omega\Big\{-\sin\left(\frac{k}{2}\omega\theta_1\right)\cos\big[k\omega t + \varphi_l(k\omega)\big]$$
$$+ \cos\left(\frac{k}{2}\omega\theta_1\right)\sin\big[k\omega t + \varphi_l(k\omega)\big]\Big\}A_l(k\omega)\Big\}. \tag{5.73}$$

In the point $\theta_1 = \pi/\omega$ this derivative becomes:

$$\frac{\partial y(t)}{\partial \theta_1} = \frac{\partial y_0}{\partial \theta_1} + \frac{4h}{\pi} \sum_{k=1}^{\infty} \frac{1}{k} \Big\{\frac{k}{2}\omega \cos\left(\frac{k\pi}{2}\right)\Big\{\cos\left(\frac{k\pi}{2}\right)\cos\big[k\omega t + \varphi_l(k\omega)\big]$$
$$+ \sin\left(\frac{k\pi}{2}\right)\sin\big[k\omega t + \varphi_l(k\omega)\big]\Big\}A_l(k\omega)$$
$$+ \sin\left(\frac{k\pi}{2}\right)\frac{k}{2}\omega\Big\{-\sin\left(\frac{k\pi}{2}\right)\cos\big[k\omega t + \varphi_l(k\omega)\big]$$
$$+ \cos\left(\frac{k\pi}{2}\right)\sin\big[k\omega t + \varphi_l(k\omega)\big]\Big\}A_l(k\omega)\Big\}, \tag{5.74}$$

or, equivalently,

$$
\frac{\partial y(t)}{\partial \theta_1} = \frac{\partial y_0}{\partial \theta_1} + \frac{4h}{\pi} \sum_{k=1}^{\infty} \frac{1}{k} \left\{ \frac{k}{2} \omega \cos^2\left(\frac{k\pi}{2}\right) \cos\left[k\omega t + \varphi_l(k\omega)\right] A_l(k\omega) \right.
$$
$$
\left. - \frac{k}{2} \omega \sin^2\left(\frac{k\pi}{2}\right) \cos\left[k\omega t + \varphi_l(k\omega)\right] A_l(k\omega) \right\}. \tag{5.75}
$$

Formula (5.75) can be transformed into:

$$
\frac{\partial y(t)}{\partial \theta_1} = \frac{\partial y_0}{\partial \theta_1} + \frac{2h}{\pi} \sum_{k=1}^{\infty} \omega \left[\cos^2\left(\frac{k\pi}{2}\right) - \sin^2\left(\frac{k\pi}{2}\right)\right] \cos\left[k\omega t + \varphi_l(k\omega)\right] A_l(k\omega),
$$
$$
\tag{5.76}
$$

which leads to the following expression when we expand the cosine of a sum:

$$
\frac{\partial y(t)}{\partial \theta_1} = \frac{\partial y_0}{\partial \theta_1} + \frac{2h}{\pi} \sum_{k=1}^{\infty} \omega \left[\cos^2\left(\frac{k\pi}{2}\right) - \sin^2\left(\frac{k\pi}{2}\right)\right] \left[\cos(k\omega t)\cos\varphi_l(k\omega)\right.
$$
$$
\left. - \sin(k\omega t)\sin\varphi_l(k\omega)\right] A_l(k\omega). \tag{5.77}
$$

For k even the term in brackets is 1, and for k odd it is -1. Therefore, the previous expression can be rewritten as

$$
\frac{\partial y(t)}{\partial \theta_1} = \frac{\partial y_0}{\partial \theta_1} + \frac{2h}{\pi} \sum_{k=1}^{\infty} \omega(-1)^k \left[\cos(k\omega t)\cos\varphi_l(k\omega)\right.
$$
$$
\left. - \sin(k\omega t)\sin\varphi_l(k\omega)\right] A_l(k\omega). \tag{5.78}
$$

Now we can compare the derivative values for time $t = t_m$ and time $t = \pi/\omega + t_m$, where t_m is the time of maximum of $y(t)$. It follows from (5.78) that the derivative at time $t = \pi/\omega + t_m$ is

$$
\left.\frac{\partial y(t)}{\partial \theta_1}\right|_{t=\pi/\omega+t_m} = \frac{\partial y_0}{\partial \theta_1} + \frac{2h}{\pi} \sum_{k=1}^{\infty} \omega(-1)^k \left[\cos\big(k\omega(t_m + \pi/\omega)\big)\cos\varphi_l(k\omega)\right.
$$
$$
\left. - \sin\big(k\omega(t_m + \pi/\omega)\big)\sin\varphi_l(k\omega)\right] A_l(k\omega) \tag{5.79}
$$

or, after applying sine/cosine to the sum formula, this can be rewritten as follows:

$$
\left.\frac{\partial y(t)}{\partial \theta_1}\right|_{t=\pi/\omega+t_m} = \frac{\partial y_0}{\partial \theta_1} + \frac{2h}{\pi} \sum_{k=1}^{\infty} \omega(-1)^k \big[\big(\cos(k\omega t_m)\cos(k\pi)
$$
$$
- \sin(k\omega t_m)\sin(k\pi)\big)\cos\varphi_l(k\omega)
$$
$$
- \big(\sin(k\omega t_m)\cos(k\pi)
$$
$$
+ \cos(k\omega t_m)\sin(k\pi)\big)\sin\varphi_l(k\omega)\big] A_l(k\omega). \tag{5.80}
$$

We note that for every k (odd and even) $\sin(k\pi) = 0$. And, therefore, (5.80) can be reduced to

$$\left.\frac{\partial y(t)}{\partial \theta_1}\right|_{t=\pi/\omega+t_m} = \frac{\partial y_0}{\partial \theta_1} + \frac{2h}{\pi}\sum_{k=1}^{\infty}\omega(-1)^k\big[\cos(k\omega t_m)\cos(k\pi)\cos\varphi_l(k\omega)$$

$$-\sin(k\omega t_m)\cos(k\pi)\sin\varphi_l(k\omega)\big]A_l(k\omega), \qquad (5.81)$$

or, equivalently, to

$$\left.\frac{\partial y(t)}{\partial \theta_1}\right|_{t=\pi/\omega+t_m} = \frac{\partial y_0}{\partial \theta_1} + \frac{2h}{\pi}\sum_{k=1}^{\infty}\omega(-1)^k\cos(k\pi)\big[\cos(k\omega t_m)\cos\varphi_l(k\omega)$$

$$-\sin(k\omega t_m)\sin\varphi_l(k\omega)\big]A_l(k\omega)$$

$$= \frac{\partial y_0}{\partial \theta_1} + \frac{2h}{\pi}\sum_{k=1}^{\infty}\omega(-1)^k(-1)^k\big[\cos(k\omega t_m)\cos\varphi_l(k\omega)$$

$$-\sin(k\omega t_m)\sin\varphi_l(k\omega)\big]A_l(k\omega)$$

$$= \frac{\partial y_0}{\partial \theta_1} + \frac{2h}{\pi}\sum_{k=1}^{\infty}\omega\big[\cos(k\omega t_m)\cos\varphi_l(k\omega)$$

$$-\sin(k\omega t_m)\sin\varphi_l(k\omega)\big]A_l(k\omega). \qquad (5.82)$$

The same derivative taken at time $t = t_m$ can be directly obtained from (5.78):

$$\left.\frac{\partial y(t)}{\partial \theta_1}\right|_{t=t_m} = \frac{\partial y_0}{\partial \theta_1} + \frac{2h}{\pi}\sum_{k=1}^{\infty}\omega(-1)^k\big[\cos(k\omega t_m)\cos\varphi_l(k\omega)$$

$$-\sin(k\omega t_m)\sin\varphi_l(k\omega)\big]A_l(k\omega). \qquad (5.83)$$

Comparing (5.82) and (5.83), one can see that positive and negative parts of the process output signal respond at a different rate to the changes of the pulse width (t_m is considered an arbitrary value).

The derivatives of the positive and negative amplitudes can be determined as follows:

$$\frac{\partial a_p}{\partial \theta_1} = \left.\frac{\partial y(t)}{\partial \theta_1}\right|_{t=t_m} + \left.\frac{\partial y(t)}{\partial t}\right|_{t=t_m}\frac{dt_m}{d\theta_1}.$$

Considering that $\frac{\partial y(t)}{\partial t}|_{t=t_m} = 0$ (where t_m is both time of maximum and time of minimum) we can write for the amplitude of the positive half-wave:

$$\frac{\partial a_p}{\partial \theta_1} = \frac{\partial y_0}{\partial \theta_1} + \frac{2h}{\pi}\omega\sum_{k=1}^{\infty}(-1)^k\big[\cos(k\omega t_{\max})\cos\varphi_l(k\omega)$$

$$-\sin(k\omega t_{\max})\sin\varphi_l(k\omega)\big]A_l(k\omega), \qquad (5.84)$$

and for the amplitude of the negative half-wave the following holds:

$$\frac{\partial a_n}{\partial \theta_1} = \frac{\partial y_0}{\partial \theta_1} + \frac{2h}{\pi}\omega\sum_{k=1}^{\infty}\big[\cos(k\omega t_{\max})\cos\varphi_l(k\omega)$$

$$-\sin(k\omega t_{\max})\sin\varphi_l(k\omega)\big]A_l(k\omega). \qquad (5.85)$$

In (5.84) and (5.85) the considered time instant is the same: $t_m = t_{max}$ and $a_n = y(t_{min})$. Considering formula (5.67) we now obtain the derivative for the shift:

$$\frac{\partial \Delta_e}{\partial \theta_1} = 0.5\beta\left(\frac{\partial a_p}{\partial \theta_1} + \frac{\partial a_n}{\partial \theta_1} - 2\frac{dr_0}{d\theta_1}\right),$$
(5.86)

where the derivatives in parentheses are given by (5.84) and (5.85).

We can compute the derivative of the shift as follows.

$$\frac{\partial \Delta_e}{\partial \theta_1} = \beta\left\{-\frac{dr_0}{d\theta_1} + \frac{\partial y_0}{\partial \theta_1} + \frac{2h}{\pi}\omega\sum_{k=1}^{\infty}[\cos(2k\omega t_{max})\cos\varphi_l(2k\omega)\right.$$

$$\left. - \sin(2k\omega t_{max})\sin\varphi_l(2k\omega)]A_l(2k\omega)\right\}$$

$$= \beta\left\{-\frac{dr_0}{d\theta_1} + \frac{\partial y_0}{\partial \theta_1} + \frac{2h}{\pi}\omega\sum_{k=1}^{\infty}[\cos(2k\omega t_{max} + \varphi_l(2k\omega))]A_l(2k\omega)\right\}.$$
(5.87)

The derivative was found above with respect to θ_1, and now we find this derivative with respect to u_0. Considering that $u_0 = h\frac{\theta_1 - \theta_2}{\theta_1 + \theta_2}$ and that the frequency is constant, the following holds: $u_0 = h(\frac{\theta_1\omega}{\pi} - 1)$. Therefore, the derivative is $\frac{d\theta_1}{du_0} = \frac{\pi}{h\omega}$. Also, it is obvious that $\frac{dy_0}{du_0} = W_l(j0)$, which results in

$$\frac{\partial \Delta_e}{\partial u_0} = \beta\left\{-\frac{dr_0}{du_0} + W_l(j0) + 2\sum_{k=1}^{\infty}[\cos(2k\omega t_{max} + \varphi_l(2k\omega))]A_l(2k\omega)\right\}. \quad (5.88)$$

Formula (5.88) includes the component $\frac{dr_0}{du_0}$ which can only be determined if the equivalent gain of the MRFT algorithm is known. Yet, it is the variable that we aim to determine. Therefore, computing the derivative (5.88) involves solving an equation for k_n. This component of (5.88) is given as follows:

$$\frac{dr_0}{du_0} = \frac{1}{k_n} + W_l(j0).$$

Denote -0.5 of the series on the right-hand side of formula (5.88) multiplied by the factor $-1/2$ as $R(\omega)$:

$$R(\omega) = \sum_{k=1}^{\infty}[\cos(2k\omega t_{max} + \varphi_l(2k\omega))]A_l(2k\omega).$$
(5.89)

Function $R(\omega)$ accounts for the unequal response of the amplitudes of the negative and positive half-waves to the change of u_0. It contains only even harmonics of the fundamental frequency component. It is a function of the process model and can be computed from the transfer function of the process. Therefore, formula (5.88) can be rewritten as follows:

$$\frac{\partial \Delta_e}{\partial u_0} = -\frac{\beta}{k_n} + 2\beta R(\Omega),$$
(5.90)

where Ω is the frequency of the oscillations. The equivalent gain of the hysteretic relay computed as per the LPRS method is $k^*_{nLPRS} = \frac{1}{-2\mathrm{Re}J(\Omega)}$, where $J(\omega)$ is the LPRS. Now, write an equation from which the formula of the equivalent gain of the MRFT algorithm can be found. It accounts for the effect of the shift of the asymmetric hysteretic relay characteristic by Δ_e as the equivalent gain of the relay k^*_{nLPRS} having a feedback with gain $\frac{\partial \Delta_e}{\partial u_0}$ (formula (5.90)):

$$k_n = \frac{k^*_{nLPRS}}{1 + k^*_{nLPRS}\beta\left(\frac{1}{k_n} - 2R(\Omega)\right)}. \tag{5.91}$$

Equation (5.91) can be solved for k_n and the formula of the equivalent gain of the MRFT algorithm can be found as follows:

$$k_n = \frac{(1-\beta)k^*_{nLPRS}}{1 - 2\beta k^*_{nLPRS}R(\Omega)}. \tag{5.92}$$

Analysis shows that the term in the denominator of (5.92) $2\beta R(\Omega)$ is small (if the oscillation is harmonic it becomes equal to zero). In this case the value of the equivalent gain computed per (5.92) would be approximately equal to the equivalent gain value computed per (5.67) (the difference would be due to the difference between the values of k^*_{nLPRS} and k^*_{nDF}). Formula (5.92) can be rewritten through LPRS notation as follows:

$$k_n = \frac{1-\beta}{-2\mathrm{Re}J(\Omega) - 2\beta R(\Omega)}. \tag{5.93}$$

Since $\mathrm{Re}\,J(\Omega)$ is normally a negative value, the plot $-2\,\mathrm{Re}\,J(\Omega) - 2\beta R(\Omega)$ can be seen as a small offset of the LPRS in either the positive or negative direction. Yet, the main change of the equivalent gain value (in comparison with the relay control) is due to the multiplier $(1-\beta)$.

When analysing the system response to an external input signal we assumed that $\frac{\partial \Omega}{\partial u_0}\big|_{u_0=0} = 0$. This property is a result of the symmetry principle: Both positive and negative changes of u_0 around $u_0 = 0$ result in the same changes of the oscillation frequency Ω. Therefore, the derivative of the frequency is zero: $\frac{\partial \Omega}{\partial u_0}\big|_{u_0=0} = 0$.

Summing up the equivalent gain derivations, we can come up with the LPRS formula of the MRFT algorithm as follows:

$$J_{MRFT}(\omega) = \frac{1}{1-\beta}\,\mathrm{Re}\,J(\omega) + \frac{\beta}{1-\beta}R(\omega) + j\,\mathrm{Im}\,J(\omega). \tag{5.94}$$

It should be noted though that the periodic solution is found not through the use of the imaginary part of J_{MRFT} but through the use of the function $\Phi(\omega)$, which in turn includes $\mathrm{Im}\,J(\omega)$. LPRS of MRFT can be used for analysis of the algorithm robustness due to the presence of external disturbances and nonlinearities, which cause asymmetry of the oscillations.

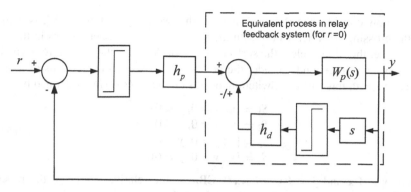

Fig. 5.13 Transformed system for two-relay test

5.4 Exact Model of Oscillations in Two-Relay Controller

The test that uses two relays for identification of process parameters was proposed in [38] and [29]—with inclusion of the integral and the derivative of the output variable, respectively. These two tests were considered in Chap. 2. It should be noted that from the perspective of the analysis of the oscillations, these two systems are equivalent. Indeed, if in the system with the *two-relay control* (Fig. 2.9) we consider the transfer function of the process $\frac{1}{s} W_p(s)$ we obtain the *two-channel relay* system, as seen in Fig. 2.8. Therefore, we develop the model of the oscillations using the model of the *two-relay controller* (Fig. 2.9), having in mind though the possibility of using this model for the two-channel relay test, too.

The exact model of oscillations in the two-relay controller was presented in [18] and further developed in [2] and [1]. It is provided below.

The system depicted in Fig. 2.9 for zero set point can be equivalently presented as a conventional relay system, which has the second relay transposed to the feedback and attributed to the process as shown in Fig. 5.13.

The transformed system can now be viewed as the relay feedback system with a nonlinear process and analysed using the LPRS concept. The computing formulas derived for linear processes cannot be directly used for LPRS computing in this case. However, considering the fact that the output of the process in a steady periodic motion is the result of the propagation of the two square pulse signals through linear dynamics, we can find the output signal using the superposition principle.

The two-relay control is given by the expression

$$u = -h_p \, \text{sign}(y) \pm h_d \, \text{sign}(\dot{y}), \tag{5.95}$$

where h_p and h_d are the amplitudes of the two relays. Consider the periodic motion in the system with the two-relay control and process given by (5.1).

We account for the presence of delay by using the Pade approximation of appropriate order. With this approach the system matrices $\mathbf{A}$, $\mathbf{B}$ and $\mathbf{C}$ will include the transfer function of the Pade approximation, which will allow us to use the results of paper [1]. It will be shown later that exact treatment of the delay can be used, too, with the approach involving the Fourier series expansion.

We now consider the Poincaré map for the two-relay system, which is defined by the crossing of either y or $\dot{y}$ with the hyperplanes giving zero levels in the state space. We, therefore, select the surface of section S in the state space R^n (where n is the system order) and consider the points of successive intersections of a given trajectory with this surface. Switching occurs on the level surfaces defined by

$$
\begin{aligned}
S_1 &= \{\mathbf{x} : y = 0, \dot{y} < 0\}, \\
S_2 &= \{\mathbf{x} : y < 0, \dot{y} = 0\}, \\
S_3 &= \{\mathbf{x} : y = 0, \dot{y} > 0\}, \\
S_4 &= \{\mathbf{x} : y > 0, \dot{y} = 0\},
\end{aligned}
\tag{5.96}
$$

where $y = \mathbf{Cx}$ and $\dot{y} = \mathbf{C\dot{x}} = \mathbf{CAx} + \mathbf{CB}u$. If the relative degree of the process model is higher than 1 then $\mathbf{CB} = 0$ and $\dot{y} = \mathbf{CAx}$.

The system is governed by one of the four models defined by

$$
\begin{aligned}
M_1 &: \dot{\mathbf{x}} = \mathbf{Ax} + \mathbf{B}(h_p + h_d), \\
M_2 &: \dot{\mathbf{x}} = \mathbf{Ax} + \mathbf{B}(h_p - h_d), \\
M_3 &: \dot{\mathbf{x}} = \mathbf{Ax} - \mathbf{B}(h_p + h_d), \\
M_4 &: \dot{\mathbf{x}} = \mathbf{Ax} + \mathbf{B}(-h_p + h_d).
\end{aligned}
$$

Switching between the models is done in accordance with the above state space partitioning and the delay between the time of entry of the trajectory into a particular partition and the selection of corresponding control. We denote the following values of the state vector x for the control switching in the periodic motion: $\rho = \mathbf{x}(0)$ corresponding to the switching to M_1, $\eta = \mathbf{x}(0)$ corresponding to the switching to M_2, $\rho^- = \mathbf{x}(0)$ corresponding to the switching to M_3, and $\eta^- = \mathbf{x}(0)$ corresponding to the switching to M_4.

We assume that the initial conditions are given by the state vector $\mathbf{x}(0) = \rho_p$, where $(\cdot)_p$ stands for "periodic", with the output and its derivative being

$$
\begin{aligned}
y(0) &= \mathbf{Cx}(0) = \mathbf{C}\rho_p = 0, \\
\dot{y}(0) &= \mathbf{C}\big(\mathbf{Ax}(0) + \mathbf{B}u\big) = \mathbf{CA}\rho_p < 0.
\end{aligned}
\tag{5.97}
$$

The solution of M_1 on the time interval $[0, t_1]$, where t_1 is the transition time from S_1 to S_2, is given by

$$
\mathbf{x}(t) = e^{\mathbf{A}t}\mathbf{x}(0) + \mathbf{A}^{-1}\big(e^{\mathbf{A}t} - \mathbf{I}\big)\mathbf{B}u(t),
$$

where $u = h_p + h_d$. The transition to S_2 and subsequent switching to $u = h_p - h_d$ is ensured under the condition of *proper direction of change*

$$
\ddot{y}(t_1) = \mathbf{CA}^2\eta_k > 0.
\tag{5.98}
$$

Analogously, for the subsequent switchings on the $(y, \dot{y})$-plane, the four point mappings, with the first beginning in $\rho_k = \rho_p$, will be governed by

$$
\begin{aligned}
\eta_k &= e^{\mathbf{A}t_1}\rho_k + \mathbf{A}^{-1}\big(e^{\mathbf{A}t_1} - \mathbf{I}\big)\mathbf{B}(h_p + h_d), \\
\rho_k^- &= e^{\mathbf{A}t_2}\eta_k + \mathbf{A}^{-1}\big(e^{\mathbf{A}t_2} - \mathbf{I}\big)\mathbf{B}(h_p - h_d), \\
\eta_k^- &= e^{\mathbf{A}t_3}\rho_k^- - \mathbf{A}^{-1}\big(e^{\mathbf{A}t_3} - \mathbf{I}\big)\mathbf{B}(h_p + h_d), \\
\rho_{k+1} &= e^{\mathbf{A}t_4}\eta_k^- - \mathbf{A}^{-1}\big(e^{\mathbf{A}t_4} - \mathbf{I}\big)\mathbf{B}(h_p - h_d),
\end{aligned}
\tag{5.99}
$$

where t_2 is the time interval between S_2 and S_3, t_3 is the time interval between S_3 and S_4, and t_4 is the time interval between S_4 and S_1.

The fixed point of the Poincaré map, corresponding to a periodic solution of the system (5.1), (5.95) driven by the two-relay controller, is determined by equation $\rho_{k+1} = \rho_k = \rho_p$. Skipping the sequential numbers of switching in (5.99) and using the principle of symmetry of the oscillation we write the following: $\rho_p^- = -\rho_p$. For the T-periodic symmetric solution we will use the following notation: $t_1 = t_3 = \theta_1$, $t_2 = t_4 = \theta_2 = T/2 - \theta_1$.

The equation for the fixed point can be rewritten as follows:

$$\eta_p = e^{\mathbf{A}\theta_1}\rho_p + \mathbf{A}^{-1}\left(e^{\mathbf{A}\theta_1} - \mathbf{I}\right)\mathbf{B}(h_p + h_d),$$
$$-\rho_p = e^{\mathbf{A}\theta_2}\eta_p + \mathbf{A}^{-1}\left(e^{\mathbf{A}\theta_2} - \mathbf{I}\right)\mathbf{B}(h_p - h_d), \tag{5.100}$$

in which we denote the variables as satisfying $y(0) = \dot{y}(\theta_1) = 0$ and $\mathbf{CB} = 0$, which in turn results in

$$\mathbf{C}\rho_p = 0, \qquad \mathbf{CA}\eta_p = 0, \qquad \mathbf{CA}\rho_p < 0, \qquad \mathbf{CA}^2\eta_p > 0.$$

We assume in (5.100) that there are no additional switches on intervals $t \in (0, \theta_1)$ and $t \in (\theta_1, \theta_1 + \theta_2)$, and the oscillation gives a *unimodal limit cycle*. This condition can be easily verified once parameters θ_1 and θ_2 are determined.

System (5.100) can be considered a system of algebraic equations for θ_1 and θ_2 if we consider the problem of analysis of the system controlled by the two-relay algorithm. Alternatively, it can be considered the set of equations for some two unknown parameters of the process in the identification problem.

Taking into account that the amplitude of the oscillations a_0 is

$$a_0 = y(\theta_1) = \mathbf{C}\eta_p \quad \text{and} \quad \theta_1 + \theta_2 = \pi/\Omega_0 = T/2 \tag{5.101}$$

in the problem of identification with measurement of a_0 and Ω_0, equations (5.100) and (5.101) can be considered a set of two nonlinear algebraic equations with respect to two unknown parameters of the process.

Note however that (5.100) and (5.101) is a system of nonlinear algebraic equations and might be hard to solve.

The following turns out to be true. The linearity of the process and the fact that the control in the periodic motion can be represented as a sum of two relay controls, wherein the response of the process is obtained as a linear combination (sum) of the two periodic relay controls of amplitudes h_p and h_d, allows for a reduction of complexity of the original problem. Let us develop an approach that might simplify finding fixed points of the Poincaré map utilising the concepts of the LPRS method.

5.4.1 LPRS-Based Analysis

Computing the LPRS of the system would allow one to easily find parameters of the periodic motion. Despite the fact that, in the problem of identification, parameters of the oscillations are measured from the test and are, therefore, known treating the frequency and amplitude of oscillations as unknown variables in the system with known parameters still makes sense. As the solution of some practical problems shows [29], variation of process parameters aimed at matching the frequency and amplitude of oscillations may produce an efficient identification algorithm.

The LPRS method presented earlier in this chapter is a method developed for analysis and design of relay systems. This method cannot be directly applied to the two-relay system, since the two-relay configuration assumes a four-level relay control versus two-level of the conventional relay system. However, after some modifications, the LPRS methodology can be used in this situation as well.

The control can be represented as a sum of two relay controls, and the output of the system can be considered a superposition of the system reaction to these two relay controls. Therefore, as an auxiliary step, let us find the Poincaré map and its fixed point in the system with one relay. Assume that the control is

$$u = \text{sign}(y). \tag{5.102}$$

Then for the part of the period for which $u = 1$

$$\mathbf{x}(t) = e^{\mathbf{A}t}\xi_p + \mathbf{A}^{-1}(e^{\mathbf{A}t} - \mathbf{I})\mathbf{B}, \tag{5.103}$$

where $\xi_p = \mathbf{x}(0)$ in the periodic motion.[2]

Assume that a symmetric periodic process of period T occurs in the system (5.1), (5.102). Then at time $t = T/2$ the state vector is

$$\mathbf{x}(T/2) = e^{\mathbf{A}T/2}\xi_p + \mathbf{A}^{-1}(e^{\mathbf{A}T/2} - \mathbf{I})\mathbf{B}, \tag{5.104}$$

which must be equal to $-\xi_p$ to provide a fixed point of the Poincaré map for the symmetric motion. Therefore, solution of the equation $-\xi_p = \mathbf{x}(T/2)$, where $\mathbf{x}(T/2)$ is given by (5.104), provides the fixed point:

$$\xi_p = (\mathbf{I} + e^{\mathbf{A}T/2})^{-1}\mathbf{A}^{-1}(\mathbf{I} - e^{\mathbf{A}T/2})\mathbf{B}. \tag{5.105}$$

Now introduce a function that provides the value of the system output in a periodic motion of the frequency $\omega = 2\pi/T$ at the time $t = \gamma T$, where $\gamma \in [-\frac{1}{2}, \frac{1}{2}]$, subject to the control amplitude being $\pi/4$ (this value of the amplitude, which is the ratio between the amplitude of the first harmonic of the square pulse signal and the amplitude of the pulses, is used to comply with the locus of a perturbed relay system (LPRS)). Taking into account (5.103) and (5.104), this function can be defined as follows:

$$\begin{aligned}
L(\omega, \gamma) &= \frac{\pi}{4}\mathbf{C}\{e^{\mathbf{A}\gamma T}\xi_p + \mathbf{A}^{-1}(e^{\mathbf{A}\gamma T} - \mathbf{I})\mathbf{B}\} \\
&= \frac{\pi}{4}\mathbf{C}\{e^{\mathbf{A}\gamma\frac{2\pi}{\omega}}(\mathbf{I} + e^{\mathbf{A}\frac{\pi}{\omega}})^{-1}\mathbf{A}^{-1}(\mathbf{I} - e^{\mathbf{A}\frac{\pi}{\omega}}) \\
&\quad + \mathbf{A}^{-1}(e^{\mathbf{A}\gamma\frac{2\pi}{\omega}} - \mathbf{I})\}\mathbf{B}.
\end{aligned} \tag{5.106}$$

---

[2]We use different notation for $\mathbf{x}(0)$ from the previous one because we consider a different type of control, which results in a different periodic motion.

Parameter γ is related to θ_1 and θ_2 in the following way:

$$\theta_1 = \gamma T$$

and

$$\theta_2 = T/2 - \theta_1 = (0.5 - \gamma)T.$$

Now consider periodic control $u(t)$ as a sum of two periodic square pulse controls $u_1(t)$ and $u_2(t)$ of amplitudes h_p and h_d, respectively. Assume that control $u_2(t)$ leads with respect to $u_1(t)$ by time $t = \gamma T$, where $\gamma \in [-0.5, 0.5]$. Then for the system output $y(t)$ at the time of the switch from $-h_p$ to $+h_p$ of the control $u_1(t)$, with $y(t)$ being the system response to the periodic control $u(t)$ of frequency Ω, we can write the following formula, which is a superposition of the responses to the two controls:

$$y(0) = \frac{4h_p}{\pi} L(\Omega, 0) + \frac{4h_d}{\pi} L(\Omega, \gamma). \tag{5.107}$$

In the exact same way, we can write the formula for the system derivative output at the time of the switch from $-h_d$ to $+h_d$ of the control $u_2(t)$:

$$\dot{y}(-\gamma T) = \frac{4h_p}{\pi} L_1(\Omega, -\gamma) + \frac{4h_d}{\pi} L_1(\Omega, 0), \tag{5.108}$$

where function L_1 would correspond to the linear plant for the output being the derivative of $y(t)$ and given by $\dot{y} = \mathbf{Cx} = \mathbf{CAx}$ (we assume that relative degree of the process model is higher than one and, therefore, $\mathbf{CB} = 0$).

Considering the equations of the closed-loop system (5.1), (5.95), one notices that the condition $y(0) = 0$ represents the switching condition for the first relay and the condition $\dot{y}(-\gamma T) = 0$ represents the switching condition for the second relay. Therefore, the fixed point of the Poincaré map for the system (5.1), (5.102) can be written as a set of two algebraic equations with two unknowns Ω and γ as follows:

$$h_p L(\Omega, 0) + h_d L(\Omega, \gamma) = 0, \tag{5.109}$$

$$h_p L_1(\Omega, -\gamma) + h_d L_1(\Omega, 0) = 0. \tag{5.110}$$

Representation of the periodic solution in the format of the LPRS can simplify the solution of equations (5.109), (5.110). This simplification comes from the consideration that the feedback through $\dot{y}(t)$ is closed and the feedback through $y(t)$ is open, thus giving a SISO plant. Finding response of this plant to the periodic discontinuous control in a certain frequency range is a simpler task. A methodology of analysis similar to that of [18] can now be used. With this approach, at the step of computing of LPRS, the frequency Ω is known, which reduces the problem to the solution of one nonlinear algebraic equation for γ. At the second step, after LPRS is computed, the actual frequency Ω is determined via finding the point of intersection of the LPRS with the real axis. Considering the definition of LPRS, at any given ω the imaginary part of the LPRS can be written in the following way:

$$\mathrm{Im}\, J(\omega) = L(\omega, 0) + \frac{h_d}{h_p} L(\omega, \gamma). \tag{5.111}$$

The value of γ in (5.111) is found from equations (5.109), (5.110), which are reduced to one equation,

$$\Upsilon(\gamma) = L(\omega, 0)L_1(\omega, -\gamma) - L(\omega, \gamma)L_1(\omega, 0) = 0, \qquad (5.112)$$

which can be solved via simple numeric algorithms.

In the present analysis, the real part of LPRS is not used in calculations, as it reflects the transfer properties of relay feedback systems [16]. The LPRS analysis of the system would include the steps of finding the value of parameter γ and computing the LPRS point for every frequency ω from the range of interest, plotting the LPRS in the complex plane and finding the point of its intersection with the real axis.

Since function $L(\omega, \gamma)$ provides the value of the system output in a periodic motion at time γT, finding the amplitude of the oscillations is equivalent to finding the maximum of L as follows:

$$a_0 = \max_{t \in [0, T]} \left\{ \frac{4h_p}{\pi} L(\Omega, t/T) + \frac{4h_d}{\pi} L(\Omega, \gamma + t/T) \right\}. \qquad (5.113)$$

However, the problem of finding the amplitude can be simplified if instead of the true amplitude given by (5.113) the amplitude of the fundamental frequency (i.e., first harmonic) can be used. In this case, using the rotating phasor concept, the control can be represented as a sum of two rotating vectors having amplitudes $4h_p/\pi$ and $4h_d/\pi$, with the angle $2\pi\gamma$ between them. The amplitude of the control vector will be

$$a_u \approx \frac{4}{\pi} \sqrt{h_p^2 + h_d^2 + 2h_p h_d \cos(2\pi\gamma)} \qquad (5.114)$$

and the amplitude of the output (taking into account only the first harmonic) will be

$$a_0 \approx \frac{4}{\pi} \sqrt{h_p^2 + h_d^2 + 2h_p h_d \cos(2\pi\gamma)} |W_p(\Omega)|, \qquad (5.115)$$

where Ω is the frequency of the periodic motion and $W_p(s) = \mathbf{C}(s\mathbf{I} - \mathbf{A})^{-1}\mathbf{B}$ is the transfer function of the process. It should be noted that this approximation based on the first harmonic is more accurate than the standard describing function approach because the frequency of the oscillations is computed exactly.

An alternative format of the L-function was proposed in [29]. It involves an infinite series of the process frequency response:

$$L(\omega, \gamma) = \sum_{k=1}^{\infty} \frac{1}{2k - 1} \left\{ \sin\left[(2k - 1)2\pi\gamma\right] \cdot \operatorname{Re} W_p\left[(2k - 1)\omega\right] \right.$$
$$\left. + \cos\left[(2k - 1)2\pi\gamma\right] \cdot \operatorname{Im} W_p\left[(2k - 1)\omega\right] \right\}. \qquad (5.116)$$

The infinite series format of the L-function (5.116) removes the necessity of the approximation of the delay. Delay can be a part of the process transfer function $W_p(s)$. Formula (5.116) was derived with the use of the Fourier series approach [18, 29], and no assumptions regarding the type of process were made at its derivation.

5.4.2 Poincaré Map-Based Analysis of Orbital Stability

Let error $e(t) = -y(t)$ be a periodic signal in the unperturbed system. This signal and the time derivative of the error $\dot{e}(t)$ produce the control $u(t) = u_1(t) + u_2(t)$.

We shall use (5.99) to analyse the deviation of a trajectory initiated on the surface S_1 at $\mathbf{x}(0) = \rho_k = \rho_p + \delta_\rho$ from a periodic trajectory initiated from some ρ_p for sufficiently small initial deviations δ_ρ. Using the equation in (5.99) for the mapping $\rho_k \to \eta_k$, the equation in (5.96) for the condition of the switch, and the Taylor expansion $e^{\mathbf{A}t_1} = e^{\mathbf{A}\theta_1} + e^{\mathbf{A}\theta_1}\mathbf{A}\Delta t + O(\Delta t^2)$, where $\Delta t = t_1 - \theta_1$ is a small value, we proceed as follows:

$$\eta_k = e^{\mathbf{A}t_1}(\rho_p + \delta_\rho) + \mathbf{A}^{-1}(e^{\mathbf{A}t_1} - \mathbf{I})\mathbf{B}(h_p + h_d)$$
$$\approx \left(e^{\mathbf{A}\theta_1} + e^{\mathbf{A}\theta_1}\mathbf{A}\Delta t\right)(\rho_p + \delta_\rho)$$
$$+ \mathbf{A}^{-1}\left(e^{\mathbf{A}\theta_1} + \left(e^{\mathbf{A}\theta_1} - \mathbf{I} + \mathbf{I}\right)\mathbf{A}\Delta t - \mathbf{I}\right)\mathbf{B}(h_p + h_d)$$

so that

$$\eta_k \approx e^{\mathbf{A}\theta_1}(\delta_\rho + \mathbf{A}\delta_\rho\Delta t) + (\mathbf{I} + \mathbf{A}\Delta t)\eta_p + \mathbf{B}(h_p + h_d)\Delta t.$$

Now, since $\mathbf{CA}\eta_k = \mathbf{CA}\eta_p = 0$, premultiplying this equation by $\mathbf{CA}$, which results in

$$\mathbf{CA}e^{\mathbf{A}\theta_1}(\delta_\rho + \mathbf{A}\delta_\rho\Delta t) + \mathbf{CA}\left(\mathbf{A}\eta_p + \mathbf{B}(h_p + h_d)\right)\Delta t \approx 0,$$

allows us to immediately obtain an estimate for $t_1 = \theta_1 + \Delta t$, which can be substituted back:

$$\eta_k = \eta_p + \delta_\eta \approx \eta_p + \varphi_1\delta_\rho,$$

where

$$\varphi_1 = \left(\mathbf{I} - \frac{v_1\mathbf{CA}}{\mathbf{CA}v_1}\right)e^{\mathbf{A}\theta_1}, \qquad v_1 = \mathbf{A}\eta_p + \mathbf{B}(h_p + h_d). \qquad (5.117)$$

Following the second equation in (5.99) and computing t_2 using $\mathbf{C}\rho_k^- = \mathbf{C}\rho_p = 0$, one, in a similar way, obtains

$$\rho_k^- = -\rho_p + \delta_{\rho-} \approx -\rho_p + \varphi_2\delta_\eta,$$

where

$$\varphi_2 = \left(\mathbf{I} - \frac{v_2\mathbf{C}}{\mathbf{C}v_2}\right)e^{\mathbf{A}\theta_2}, \qquad v_2 = \mathbf{A}\rho_p + \mathbf{B}(h_p - h_d). \qquad (5.118)$$

Following the third equation in (5.99) and computing t_3 using $\mathbf{CA}\eta_k^- = \mathbf{CA}\eta_p = 0$, one obtains

$$\eta_k^- = -\eta_p + \delta_{\eta-} \approx -\eta_p + \varphi_3\delta_{\rho-},$$

where $\varphi_3 = \varphi_1$.

Following the last equation in (5.99) and computing t_4 using $\mathbf{C}\rho_{k+1} = \mathbf{C}\rho_p = 0$, one obtains

$$\rho_{k+1} \approx \rho_p + \varphi_4\delta_{\rho-},$$

where $\varphi_4 = \varphi_2$.

Finally, we have for small $\delta_\rho = \rho_k - \rho_p$: $\rho_{k+1} - \rho_p = \Phi \cdot (\rho_k - \rho_p) + O(\delta_\rho^2)$, with

$$\Phi = (\varphi_2 \cdot \varphi_1)^2. \tag{5.119}$$

Since we have just computed a linearisation for the Poincaré map, we conclude with the following theorem, which was presented in [1].

Theorem 5.1 *Suppose that the two-relay control (5.95) induces a periodic trajectory in the closed-loop system with the process given by (5.1), where $\tau = 0$. This solution is orbitally exponentially stable if and only if all eigenvalues of the matrix Φ, defined by (5.119), (5.117) and (5.118), are located inside the unit circle.*

5.5 Example of Identification

Consider an example of identification through the two-relay control, which was presented in [29]. Despite being from the area of electrical engineering, the approach given in this example is fully applicable to process control area.

The electrical circuit that contains resistors, capacitors and operational amplifiers, with precisely measured values of the component parameters, has the following nominal transfer function:

$$W_{nom} = \frac{1.024}{(0.0096s + 1)(0.0136s + 1)(0.0194s + 1)}. \tag{5.120}$$

Identification of the parameters of the model of the circuit was done using two tests through the two-relay control presented above, with parameters $h_p = 5$ V and $h_d = 0$ V for the first test, and $h_p = 5$ V and $h_d = -1$ V for the second test. The experimentally measured frequency and the amplitude of the oscillations are: $\Omega_1 = 125.72$ rad/s and $a_1 = 0.78$ V for the first test, and $\Omega_2 = 113.09$ rad/s and $a_2 = 0.96$ V for the second test. The underlying model of the circuit was the third order transfer function of the structure given by (5.120).

Identification was done using the LPRS model presented above (formulas (5.109)–(5.113)) via the solution of four nonlinear algebraic equations for the four unknown parameters. The solution was carried out via the Newton–Raphson method. It took a short computing time and produced the following result of the identification:

$$W_{id} = \frac{1.024}{(0.0104s + 1)(0.0104s + 1)(0.0238s + 1)}.$$

One can see that the accuracy of identification is very high for some parameters and lower for others. Overall accuracy is quite satisfactory. The provided example through an experimental application demonstrates feasibility of the described method. An interested reader might find more details on this example in [29]. This publication also provides another example of the use of the same identification method from the area of mechatronics.

5.6 Conclusions

In this chapter, we provide an exact approach to analysis of oscillations in relay feedback systems, systems with the modified relay feedback test algorithm and systems with the two-relay control. The approach is based on the frequency-domain characteristic of the process called the *locus of a perturbed relay system* (LPRS). Formulas, algorithms and MATLAB code for LPRS computing were presented. Analysis of the effect of external disturbances on the system having oscillations and analysis of orbital asymptotic stability of the considered class of systems are presented, too. The former problem is solved with the use of the LPRS and the latter is done through the design of the linearised Poincaré map relating an initial deviation from the periodic motion at the time of the switch to deviations of the state vector from the one in the periodic motion at subsequent switches of the control. The presented theoretical development can be used for identification of process parameters from the RFT, MRFT and tests involving the two-relay control (with an additional integrator or with an additional differentiator). An example involving experimental measurements of the test oscillations generated by the two-relay control is provided.

Chapter 6
Analysis of Transient Oscillations in Systems with MRFT

Transient oscillations in relay feedback systems and other nonlinear systems are considered in this chapter. The consideration is based on the dynamic harmonic balance principle presented in the chapter. In addition to the relationship between the amplitude and the frequency of the oscillations, the dynamic harmonic balance involves the relationship between the rates of change of the amplitude and of the frequency, which provides a simple model of oscillations in their transient conditions. The provided theory is illustrated by example of analysis of transient oscillations.

6.1 Dynamic Harmonic Balance

6.1.1 Introduction

The harmonic balance (HB) principle is a convenient tool for finding parameters of self-excited periodic motions. Due to its convenience and simplicity, it is widely used in many areas of science and engineering. This principle was used in Chap. 2 for deriving the model of the oscillations in MRFT and deriving the constraints used in the tuning rules, from either the gain or the phase margin. We can see that for a system with one nonlinearity and linear dynamics (a Lure system), the HB principle can be illustrated by the Nyquist plot of the linear dynamics and the plot of the negative reciprocal of the describing function (DF) [8] of the nonlinearity in the complex plane and finding the point of intersection of the two plots, which would correspond to the self-excited periodic motion in the system. Therefore, the harmonic balance principle treats the system as a loop connection of the linear dynamics and of the nonlinearity. It is also possible to reformulate the harmonic balance so that the format of the system analysed is not a loop connection but the denominator of the closed-loop system. This would imply a different interpretation of the harmonic balance, which would allow one to extend the harmonic balance principle to analysis of not only self-excited periodic motions but also other types of oscillatory motions.

I. Boiko, *Non-parametric Tuning of PID Controllers*, Advances in Industrial Control, 141
DOI 10.1007/978-1-4471-4465-6_6, © Springer-Verlag London 2013

Examples of vanishing oscillatory motion of variable frequency can be found in the conventional [81] and second-order sliding mode (SM) control system models [9, 49]. Another example of a transient oscillatory motion is the transient processes in oscillatory mechanical systems. A number of examples are given in [61, 67]. Examples from different areas of engineering and physics can be found where transient oscillatory motions of variable frequency occur. The problem of the convergence rate assessment, including qualitative (finite-time or asymptotic) and quantitative assessment, is of high importance for these systems.

Transient oscillations were studied in [70] (see also [84]), [39] and some recent publications [20, 26, 69].

The availability of the model of the system revealing transient oscillations might seem to be sufficient for analysis and not require development of methods other than simulations. However, the frequency-domain approach to assessment of convergence rate would provide a number of advantages over the direct solution of the system differential equations. The most important one is the possibility of explanation of the mechanism of the frequency and amplitude change during the transient. With respect to MRFT and other tests using the continuous cycling approach, the model of transient oscillations would allow one to estimate the convergence time for the periodic motion to be established in the systems. This model would allow one to compare different methods and correctly select the running time for the test.

In this chapter, a frequency-domain approach to analysis of transient oscillations is presented, which is suitable for analysis of high-order nonlinear systems. The harmonic balance principle is extended to the case of transient oscillations and named the *dynamic harmonic balance* principle [28].

At first the conventional harmonic balance principle is considered. Then a Lure system with a high-order plant is analysed with the use of the dynamic harmonic balance involving a quasi-static approach to the frequency of the oscillations. Such characteristics as frequency and amplitude of oscillations as functions of time are derived. After that the condition of the full dynamic harmonic balance is derived. Finally, the approach is applied to analysis of the transient motions in a system with MRFT.

6.1.2 Harmonic Balance for Transient Oscillations

Consider the system that includes linear dynamics given by the following equations:

$$
\begin{aligned}
\dot{\mathbf{x}} &= \mathbf{A}\mathbf{x} + \mathbf{B}u, \\
y &= \mathbf{C}\mathbf{x},
\end{aligned}
\tag{6.1}
$$

where $\mathbf{x} \in R^n$, $y \in R^1$, $u \in R^1$, $\mathbf{A} \in R^{n \times n}$, $\mathbf{B} \in R^{n \times 1}$ and $\mathbf{C} \in R^{1 \times n}$, and a single-valued odd-symmetric nonlinearity $f(y)$:

$$
u = -f(y).
\tag{6.2}
$$

We shall refer to (6.1) as the linear part of the system. One can see that system (6.1), (6.2) is a Lure system. The transfer function of the linear part is $W_l(s) =$

$\mathbf{C}(\mathbf{I}s - \mathbf{A})^{-1}\mathbf{B}$, which can also be presented as a ratio of two polynomials $W_l(s) = P(s)/Q(s)$. Assume also an autonomous mode, so that the input to the nonlinearity is the output of the linear dynamics, and the output of the nonlinearity is the input to the linear dynamics. Self-excited periodic motions in the system can be found through the use of the HB principle. The conventional HB condition (for periodic motion) is formulated as

$$W_l(j\Omega)N(a) = -1, \qquad (6.3)$$

where Ω is the frequency and a is the amplitude of the self-excited periodic motion at the input to the nonlinearity and $N(a)$ is the describing function of the nonlinearity. If the linear part has relative degree higher than *two* then the Nyquist plot of system (6.1) has a point of intersection with the real axis at some finite frequency and, therefore, equation (6.3) has a solution. We now find the closed-loop transfer function $W_{cl}(s)$ of system (6.1), (6.2) using the replacement of the nonlinearity with the describing function (DF) $u = -N(a) \cdot y$:

$$W_{cl}(s) = \frac{W_l(s)N(a)}{1 + W_l(s)N(a)} = \frac{P(s)N(a)}{Q(s) + P(s)N(a)}. \qquad (6.4)$$

Let us note that (6.3) is equivalent to:

$$R(a, j\Omega) = Q(j\Omega) + P(j\Omega)N(a) = 0, \qquad (6.5)$$

which means that the denominator of the closed-loop transfer function becomes zero when the frequency and the amplitude become equal to the frequency and the amplitude of the periodic motion. Equation (6.5) is also sometimes used for finding a periodic solution via algebraic methods. However, equation (6.3) usually is not attributed to the denominator of the closed-loop transfer function but considered a direct result of (6.3). Assuming that $R(a, s)$ can be represented in the following form $R(a, s) = (s - s_1)(s - s_2) \cdots (s - s_n)$, where s_i are roots of the characteristic polynomial, we must conclude that there must be at least one pair of complex conjugate roots with zero real parts. It would imply the existence of the conservative component in $W_{cl}(s)$. Indeed, we can consider the existence of nonvanishing oscillations as a result of the existence of the component $(s^2 + \rho^2)$ in the denominator of $W_{cl}(s)$, where ρ is a parameter that depends on the amplitude a. However, one notices that even if a damped oscillation occurs, so that there exists a pair of complex conjugate roots s_i, s_{i+1}, then $(s - s_i)(s - s_{i+1}) = 0$, and the characteristic polynomial becomes zero, with $s = \sigma \pm j\Omega$, where σ is the decay. (Note: strictly speaking, we have a decaying oscillation only if $\sigma < 0$; yet we will refer to this variable as to the decay even if $\sigma \geq 0$.)

We can view the harmonic balance condition (6.3) as the realisation of the *regeneration principle*, according to which $y(t) = L^{-1}[L[u(t)]W_l(s)]$ and $u(t) = f(y(t))$. We can now use the regeneration principle approach a transient oscillation, in which we consider both $u(t)$ and $y(t)$ sinusoidal signals with exponentially decaying (or growing) amplitude, and formulate and prove the following property.

Theorem 6.1 *With the input signal to the linear dynamics given by the transfer function $W_l(s)$ being the harmonic signal with decaying amplitude $u(t) =$*

$e^{\sigma t}\sin(\Omega t)$, *the output of the linear dynamics is also an harmonic signal of the same frequency and decay*: $y(t) = ae^{\sigma t}\sin(\Omega t + \varphi)$.

Proof It follows from the property of the Laplace transform that $L[e^{-a}f(t)] = F(s + a)$. Therefore, for the system input $u(t)$, the Laplace transform will be $L[u(t)] = \Omega/[(s - \sigma)^2 + \Omega^2]$, which will result in the system output (in the Laplace domain) $Y(s) = \Omega W_l(s)/[(s - \sigma)^2 + \Omega^2]$. The substitution $s' = s - \sigma$ yields $Y(s') = \Omega W_l(s' + \sigma)/[(s')^2 + \Omega^2]$, which means that $y'(t) = L^{-1}[Y(s')]$ is a sinusoid of frequency Ω, amplitude $|W_l(\sigma + j\Omega)|$, and having phase shift $\arg W_l(\sigma + j\Omega)$. In turn, the output signal is $y(t) = e^{\sigma t}y'(t)$, i.e., a decaying sinusoid.   □

Therefore, for our analysis of propagation of the decaying sinusoids through linear dynamics we can use the same transfer functions, in which the Laplace variable should be replaced with $(\sigma + j\Omega)$.

The describing function N in the case of a transient oscillation may become a function of not only amplitude but of its derivatives, too [39] (we disregard possible dependence of the DF on the frequency). Considering that conditions (6.3) and (6.5) are equivalent and the equality of the denominator of the closed-loop transfer function to zero (for some s) implies the fulfilment of (6.3), we can rewrite (6.3) for the transient oscillation as

$$N(a, \dot{a}, \ldots)W_l(\sigma + j\Omega) = -1. \tag{6.6}$$

The use of the derivatives of the amplitude as arguments of the DF is inconvenient because it results in the necessity of consideration of additional variables (derivatives of the amplitude) that are not present otherwise. It is more convenient to consider σ and its derivatives than the derivatives of the amplitude. Also, we limit our consideration of the describing function arguments to the first derivative of the amplitude (or, equivalently, to σ) only, which will correspond to the use of the regeneration principle for decaying sinusoids. We show below that for many nonlinearities the DF is a function of the amplitude only—as it is in the conventional DF analysis. Therefore, we can write the condition of the existence of a transient or steady oscillation as follows:

$$N(a, \sigma)W_l(\sigma + j\Omega) = -1. \tag{6.7}$$

Equation (6.7) is referred to in [25] as the *dynamic harmonic balance* condition (equation). However, it does not account for the derivative of the frequency and we shall further refer to (6.7) as *dynamic harmonic balance quasi-static w.r.t. frequency*.

Assume now that the characteristic polynomial of the closed-loop system (with parametric dependence on the amplitude of the oscillations) has a pair of complex conjugate roots with negative real parts. Then a vanishing oscillation of certain frequency and amplitude occurs. The idea of considering equations of vanishing oscillations is similar to the one of the Krylov–Bogoliubov method [48]. However, the latter can only deal with small "deviations" from the harmonic oscillator and is limited to second-order systems. In the present approach, the "equivalent damping" is not limited to small values. Let us consider instantaneous values of the frequency, amplitude and decay and formulate the *dynamic HB quasi-static w.r.t. frequency* as follows.

At every time, a single-frequency mode transient oscillation can be described as a process of variable (instantaneous) frequency, amplitude and decay that must satisfy equation (6.7).

Note: In (6.7) and the formulation given above, we consider only transient oscillations with zero mean and single-frequency mode when the characteristic polynomial (6.5) has only one pair of complex conjugate roots.

The describing function of an arbitrary nonlinearity for nonharmonic input signal should be computed as suggested in [70] (see also [84]).

$$N(a, \sigma, \omega) = q(a, \sigma, \omega) + q'(a, \sigma, \omega) \frac{s - \sigma}{\omega}, \tag{6.8}$$

where $s = \frac{d}{dt}$, $q = \frac{1}{2\pi} \int_0^{2\pi} f[a \sin \Psi, a(\sigma \sin \Psi + \omega \cos \Psi)] \sin \Psi \, d\Psi$ and $q' = \frac{1}{2\pi} \int_0^{2\pi} f[a \sin \Psi, a(\sigma \sin \Psi + \omega \cos \Psi)] \cos \Psi \, d\Psi$, which accounts for the dependence on the frequency, also. It should be noted that for single-valued or hysteretic symmetric nonlinearities, formula (6.8) produces conventional describing function expressions that are commonly used for harmonic inputs.

The overall motion can now be obtained from the dynamics HB as follows:

$$y(t) = a(t)e^{\sigma(t)t} \sin \Psi(t), \tag{6.9}$$

where $a(t)$, $\sigma(t)$ are obtained from the following differential equation:

$$\dot{a}(t) = a(t)\sigma(t), \qquad a(0) = a_0, \tag{6.10}$$

and $\Psi(t)$ is the phase computed as $\Psi(t) = \int_0^t \Omega(\tau) d\tau + \phi$, where $\Omega(t)$ is obtained from (6.7) and ϕ is selected to satisfy initial conditions.

6.2 Analysis of Motions in the Vicinity of a Periodic Solution

Carry out frequency-domain analysis of the transient process of the convergence to the periodic motion in the vicinity of a periodic solution in system (6.1), (6.2) using the dynamic harmonic balance condition (6.7). We can write the conventional harmonic balance condition, which can also be obtained from (6.7) when $\sigma = 0$, as follows:

$$N(a_0) W_l(j\Omega_0) = -1, \tag{6.11}$$

where Ω_0 and a_0 are the frequency and the amplitude of the periodic solution. Write the dynamics harmonic balance condition for the increments from the periodic solution:

$$N(a_0 + \Delta a, \sigma) W_l(\sigma + j(\Omega_0 + \Delta \Omega)) = -1. \tag{6.12}$$

We now take the derivative from both sides of (6.12) with respect to Δa (or a) in the point $a = a_0$:

$$\left(\frac{\partial N(a,\sigma)}{\partial a}\bigg|_{a=a_0} + \frac{\partial N(a,\sigma)}{\partial \sigma}\bigg|_{\sigma=0}\frac{d\sigma}{da}\bigg|_{a=a_0}\right)W_l(j\Omega_0)$$

$$+ N(a_0)\frac{dW_l(s)}{ds}\bigg|_{s=j\Omega_0}\frac{ds}{da}\bigg|_{a=a_0} = 0. \tag{6.13}$$

At first we limit our analysis only to nonlinearities having describing functions that do not depend on σ (for example, the ideal relay nonlinearity). Later the same analysis can be applied to nonlinearities that depend on σ. Express the derivative $\frac{ds}{da}|_{a=a_0}$ from equation (6.13):

$$\frac{ds}{da}\bigg|_{a=a_0} = -\frac{\frac{dN(a)}{da}|_{a=a_0}W_l(j\Omega_0)}{N(a_0)\frac{dW_l(s)}{ds}|_{s=j\Omega_0}}. \tag{6.14}$$

Considering that $s = \sigma + j\Omega$, we can rewrite equation (6.14) as follows:

$$\frac{d\sigma}{da}\bigg|_{a=a_0} + j\frac{d\Omega}{da}\bigg|_{a=a_0} = -\frac{\frac{dN(a)}{da}|_{a=a_0}W_l(j\Omega_0)}{N(a_0)\frac{dW_l(s)}{ds}|_{s=j\Omega_0}}. \tag{6.15}$$

Equation (6.15) is a complex equation. It can be split into two equations for the real and imaginary parts. However, only the real parts of (6.15) give an equation that has a solution. Once it is solved and $a(t)$ is found, $\Omega(t)$ can be found, too. Considering that

$$\frac{1}{W_l(s)}\frac{dW_l(s)}{ds}\bigg|_{s=j\Omega_0} = \frac{d\ln W_l(s)}{ds}\bigg|_{s=j\Omega_0}$$

$$= \frac{d\arg W_l(j\omega)}{d\omega}\bigg|_{\omega=j\Omega_0} - j\frac{d\ln|W_l(j\omega)|}{d\omega}\bigg|_{\omega=j\Omega_0},$$

and

$$-\frac{1}{N(a)}\frac{dN(a)}{da}\bigg|_{a=a_0} = \frac{d\ln\tilde{N}(a)}{da}\bigg|_{a=a_0} = \frac{d\ln|\tilde{N}(a)|}{da}\bigg|_{a=a_0} + j\frac{d\arg\tilde{N}(a)}{da}\bigg|_{a=a_0},$$

where $\tilde{N}(a) = -N^{-1}(a)$, we can write for the real part of (6.15):

$$\frac{d\sigma}{da} = \mathrm{Re}\left\{\frac{\frac{d\ln|\tilde{N}(a)|}{da} + j\frac{d\arg\tilde{N}(a)}{da}}{\frac{d\arg W_l(j\omega)}{d\omega} - j\frac{d\ln|W_l(j\omega)|}{d\omega}}\right\},$$

which can be rewritten as follows (we skip for brevity the notation of the point in which the derivative is taken):

$$\frac{d\sigma}{da} = \frac{\frac{d\ln|\tilde{N}(a)|}{da}\frac{d\arg W_l(j\omega)}{d\omega} - \frac{d\arg\tilde{N}(a)}{da}\frac{d\ln|W_l(j\omega)|}{d\omega}}{\left(\frac{d\arg W_l(j\omega)}{d\omega}\right)^2 + \left(\frac{d\ln|W_l(j\omega)|}{d\omega}\right)^2}. \tag{6.16}$$

As a "side" product of our analysis, stability of a periodic solution can be assessed from $\frac{d\sigma}{da}|_{a=a_0} < 0$.

6.3 Dynamic Harmonic Balance Including Frequency Rate of Change (Full Dynamic Harmonic Balance)

In many cases, such as analysis of motions in the vicinity of a periodic solution, the dynamic HB quasi-static w.r.t. to frequency derivative is quite capable of providing a precise result because the frequency changes insignificantly and the derivative of the frequency can be neglected. In some other cases, such as analysis of convergence of systems with second-order sliding modes [26] and finite-time convergence, the system experiences significant changes of the instantaneous frequency of the oscillations. In this situation the quasi-static approach to the account of the oscillation frequency may result in the loss of accuracy of analysis. Therefore, inclusion of the frequency derivative would be desirable in such cases.

Consider the following illustrative example. Let the linear part be the second-order dynamics and the controller be the ideal relay with amplitude h. The describing function of the relay is $N(a) = \frac{4h}{\pi a}$. If we assume that the input to the nonlinearity is a decaying sinusoid of constant frequency $y(t) = a_0 e^{\sigma t} \sin(\omega t)$ then the control amplitude should be $a_u = aN(a)$, the derivative of the control amplitude $\dot{a}_u = \frac{\partial N}{\partial a} \dot{a} a + N \dot{a}$ and the decay $\sigma_u = \frac{\dot{a}_u}{a_u} = \frac{-a \frac{4h}{\pi a^2} \dot{a} + \frac{4h}{\pi a} \dot{a}}{\frac{4h}{\pi a} a} = 0$. The result showing that $\sigma_u = 0$ is quite predictable because the output of the relay controller is an oscillation of constant amplitude, which features zero decay. However, it is well known (see [8, 39], for example) that the output of the linear part represents oscillations of decaying amplitude and growing frequency. Because the input to the linear dynamics has zero decay, we can conclude that the decay in the signal $y(t)$ is a result not of the decay in the control $u(t)$ but because of the variable frequency of $u(t)$. Moreover, the nonlinearity may change the decay according to

$$\sigma_u = \frac{\dot{a}_u}{a_u} = \frac{\frac{\partial N}{\partial a} \dot{a} a + N \dot{a}}{Na} = \frac{\partial \ln N}{\partial a} \dot{a} + \sigma = \left(\frac{\partial \ln N}{\partial \ln a} + 1 \right) \sigma, \qquad (6.17)$$

and this change is offset by the decay change due to the frequency variation at the signal propagation through the linear part. This situation is not covered by equation (6.7), which assumes constant frequency; it requires the development of a different model.

Another example is the propagation of a sinusoid of nearly constant amplitude and increasing frequency through the integrator. We assume that the signal of ideally constant amplitude is $u^*(t) = \cos(\omega_0 t + \frac{1}{2}\dot{\omega}_0 t^2)$, propagates through the integrator producing signal $y^*(t) \approx y(t) = \frac{1}{\omega_0 + \dot{\omega}_0 t} \sin(\omega_0 t + \frac{1}{2}\dot{\omega}_0 t^2) = \frac{1}{\omega} \sin \Psi$ (the approximate equality is valid if $\dot{\omega}_0$ is small enough). We note that the instantaneous phase of the signal is $\Psi(t) = \omega_0 t + \frac{1}{2}\dot{\omega}_0 t^2$. Therefore, the instantaneous frequency is $\omega(t) = \dot{\Psi}(t) = \omega_0 + \dot{\omega}_0 t$, and the instantaneous derivative of the frequency is $\ddot{\Psi}(t) = \dot{\omega}_0 = \text{const}$. Find the derivative of $y(t)$ as follows:

$$\dot{y}(t) = \cos\left(\omega_0 t + \frac{1}{2}\dot{\omega}_0 t^2\right) + \frac{\dot{\omega}_0}{(\omega_0 + \dot{\omega}_0 t)^2} \sin\left(\omega_0 t + \frac{1}{2}\dot{\omega}_0 t^2\right)$$

$$= \cos \Psi + \frac{\dot{\omega}_0}{\omega^2} \sin \Psi.$$

Therefore, the integrator introduces the phase lag less than $90°$ and the equations of the integrator can be represented by the conventional integrator having an additional feedback, which accounts for the effect of the variable frequency.

Therefore, the frequency response of the integrator to a decaying sinusoid of variable frequency is $W(\sigma + j\omega, \dot{\omega}_0) = \dfrac{1}{\sigma + \frac{\dot{\omega}_0}{\omega^2} + j\omega}$. From this formula, one can see that if $\sigma = -\frac{\dot{\omega}_0}{\omega^2}$ then the phase lag introduced by the integrator $\varphi = \arg W(\sigma + j\omega, \dot{\omega}_0)$ is $90°$. The amplitude characteristic of the integrator is given by $M = |W(\sigma + j\omega, \dot{\omega}_0)|$, and the decay introduced by the integrator due to variable frequency is $\sigma(t) = \frac{\dot{a}}{a} = -\frac{\dot{\omega}_0}{\omega}$ (given the amplitude of $y(t)$ being $a = \frac{1}{\omega_0 + \dot{\omega}_0 t}$).

To analyse propagation of a decaying sinusoid of variable frequency through higher-order linear dynamics we shall use the notion of the *rotating phasor*. Let the output of the linear dynamics $y(t)$ be given by

$$\bar{y}(t) = a(t)e^{j\Psi(t)} = e^{\ln a(t)}e^{j\Psi(t)} = e^{\ln a(t) + j\Psi(t)}, \tag{6.18}$$

where a represents the length of the phasor and Ψ represents the angle between the phasor and the real axis. We can associate either the real or the imaginary part of $\bar{y}(t)$ with the real signal $y(t)$. We assume that the transfer function of the linear part does not have any zeros: $W_l(s) = 1/Q(s) = 1/(a_0 + a_1 s + a_2 s^2 + \cdots + a_n s^n)$. Given this transfer function, we can write for $y(t)$:

$$u = a_0 y + a_1 \dot{y} + a_2 \ddot{y} + \cdots + a_n y^{(n)}. \tag{6.19}$$

We can find the derivatives of $y(t)$ as follows:

$$\dot{\bar{y}}(t) = e^{\ln a(t) + j\Psi(t)}\left(\frac{d\ln a}{dt} + j\dot{\Psi}\right) = e^{\ln a(t) + j\Psi(t)}\left(\frac{1}{a}\dot{a} + j\dot{\Psi}\right)$$
$$= e^{\ln a(t) + j\Psi(t)}(\sigma + j\omega) = (\sigma + j\omega)\bar{y}, \tag{6.20}$$

$$\ddot{\bar{y}}(t) = (\dot{\sigma} + j\dot{\omega})\bar{y} + (\sigma + j\omega)\dot{\bar{y}} = [(\dot{\sigma} + j\dot{\omega}) + (\sigma + j\omega)^2]\bar{y}.$$

Considering that we disregard all derivatives of the decay and higher than first derivatives of the frequency in our model we can rewrite the last formula as

$$\ddot{\bar{y}}(t) \approx [(\sigma + j\omega)^2 + j\dot{\omega}]\bar{y} \tag{6.21}$$

and the formula for the third derivative as

$$\dddot{\bar{y}}(t) \approx [2(\sigma + j\omega)(\dot{\sigma} + j\dot{\omega}) + j\ddot{\omega}]\bar{y} + [(\sigma + j\omega)^2 + j\dot{\omega}](\sigma + j\omega)\bar{y}$$
$$\approx [j3\dot{\omega} + (\sigma + j\omega)^2](\sigma + j\omega)\bar{y} = [(\sigma + j\omega)^3 + j3\dot{\omega}(\sigma + j\omega)]\bar{y}, \tag{6.22}$$
$$\bar{y}^{(4)}(t) \approx [(\sigma + j\omega)^4 + j6\dot{\omega}(\sigma + j\omega) - 3\dot{\omega}^2]\bar{y}.$$

We can continue with taking further derivatives. It is worth noting that the formulas above are organised to have a term $(\sigma + j\omega)$ to the respective power and the term which is the product of $\dot{\omega}$ and another multiplier. Therefore, we can write for $y(t)$:

$$\bar{u} = [a_0 + a_1(\sigma + j\omega) + a_2(\sigma + j\omega)^2 + \cdots + a_n(\sigma + j\omega)^n]\bar{y} + S(\sigma, \omega, \dot{\omega})\bar{y}, \tag{6.23}$$

where $S(\sigma, \omega, \dot{\omega})$ includes all terms containing $\dot{\omega}$. This component can be accounted for as an additional feedback—the same way it was done for the integrator.

If we introduce a certain modified frequency response as

$$W_l^*(\sigma, \omega, \dot{\omega}) = \frac{\bar{y}}{\bar{u}} = \frac{1}{Q(\sigma + j\omega) + S(\sigma, \omega, \dot{\omega})} = \frac{W_l(\sigma + j\omega)}{1 + W_l(\sigma + j\omega)S(\sigma, \omega, \dot{\omega})}, \tag{6.24}$$

we can write the *dynamic harmonic balance* equation as

$$N(a)W_l^*(\sigma, \omega, \dot{\omega}) = -1. \tag{6.25}$$

Obviously, equation (6.25) can be split into two equations: for real and imaginary parts, or for equations of the magnitude balance and phase balance.

Equation (6.25) must be complemented with an equation that relates the difference of the decays at the input and the output of the linear part and the frequency rate of change—the same way it was done in the example of the integrator analysis. We note that the amplitude of the control (the first harmonic) is $a_u = a|Q(\sigma + j\omega) + S(\sigma, \omega, \dot{\omega})|$, therefore, its time derivative is

$$\dot{a}_u = \dot{a}|Q(\sigma + j\omega) + S(\sigma, \omega, \dot{\omega})| + a\frac{d|Q(\sigma + j\omega) + S(\sigma, \omega, \dot{\omega})|}{dt}.$$

The decay of signal $u(t)$ is computed as

$$\sigma_u = \frac{\dot{a}_u}{a_u} = \frac{\dot{a}|Q(\sigma + j\omega) + S(\sigma, \omega, \dot{\omega})| + a\frac{d|Q(\sigma+j\omega)+S(\sigma,\omega,\dot{\omega})|}{dt}}{a|Q(\sigma + j\omega) + S(\sigma, \omega, \dot{\omega})|}$$

$$= \sigma + \frac{d\ln|Q(\sigma + j\omega) + S(\sigma, \omega, \dot{\omega})|}{dt}.$$

Because we disregard $\dot{\sigma}$ and $\ddot{\omega}$ we can rewrite the last formula as follows:

$$\sigma_u = \sigma + \frac{\partial \ln|Q(\sigma + j\omega) + S(\sigma, \omega, \dot{\omega})|}{\partial \omega}\dot{\omega} = \sigma - \frac{\partial \ln|W_l^*(\sigma, \omega, \dot{\omega})|}{\partial \omega}\dot{\omega}.$$

The derivative in the last formula defines the slope of the magnitude-frequency characteristic of $W_l^*(\sigma, \omega, \dot{\omega})$. Considering also the formula for the decay of $u(t)$ derived above through the DF, we can now write the condition of the balance of the decays in the closed-loop system as

$$\frac{\partial \ln|W_l^*(\sigma, \omega, \dot{\omega})|}{\partial \ln \omega}\frac{\dot{\omega}}{\omega} = -\frac{\partial \ln N(a)}{\partial \ln a}\sigma. \tag{6.26}$$

We can formulate the *dynamic harmonic balance (DHB) principle* as the following theorem, the proof of which is given above.

Theorem 6.2 *At every time during a single-frequency mode transient oscillation, the oscillation can be described as a process of variable instantaneous frequency, amplitude, decay and frequency rate of change (time derivative) that must satisfy equations (6.25) and (6.26).*

It is worth noting that $\frac{\partial \ln |W_l^*(\sigma, \omega, \dot{\omega})|}{\partial \ln \omega} = 0.05 \frac{dM(\omega)}{d\log \omega}$, where $M(\omega)$ is the magnitude frequency response [dB] (the Bode plot) for the transfer function $W_l^*(\sigma, \omega, \dot{\omega})$. Therefore, this term gives the slope of the Bode plot. Both $\dot{\omega}/\omega$ and $\sigma = \dot{a}/a$ are relative rates of change of the frequency and the amplitude, respectively. Equation (6.26), therefore, is establishing the balance between these two rates of change.

Overall solution is obtained via integration of the first-order differential equation (6.10) with σ expressed through a using the three algebraic equations presented above.

6.4 Model of Transient Oscillations in the Presence of Delay

We have considered the model of the transient oscillations that uses the DHB principle. We are going now to extend this model to plants/processes that have a time delay in the loop. We assume that the delay is applied to the control, which is not a limitation of the approach because of the commutativity property. Suppose that $u(t)$ is the original (undelayed) control and $u_1(t) = u(t - \tau)$ is the delayed control. Again, we first assume that

$$u(t) = \cos\left(\omega_0 t + \frac{1}{2}\dot{\omega}_0 t^2\right), \tag{6.27}$$

where $\Psi(t) = \omega_0 t + \frac{1}{2}\dot{\omega}_0 t$ is the instantaneous phase of $u(t)$. The delayed control is

$$u_1(t) = \cos\left(\omega_0(t - \tau) + \frac{1}{2}\dot{\omega}_0(t - \tau)^2\right),$$

where $\Psi_1(t) = \omega_0(t - \tau) + \frac{1}{2}\dot{\omega}_0(t - \tau)^2$ is the instantaneous phase of $u_1(t)$.

The instantaneous frequencies of $u(t)$ and $u_1(t)$ are

$$\omega(t) = \frac{d\Psi(t)}{dt} = \omega_0 + \dot{\omega}_0 t$$

and

$$\omega_1(t) = \frac{d\Psi_1(t)}{dt} = \omega_0 + \dot{\omega}_0(t - \tau) = \omega(t) - \tau\dot{\omega}_0 = \omega(t) - \tau\frac{d\omega(t)}{dt},$$

respectively. The phase shift created by the delay is

$$\Delta\Psi = \Psi_1(t) - \Psi(t) = \omega_0(t - \tau) + \frac{1}{2}\dot{\omega}_0(t - \tau)^2 - \omega_0 t - \frac{1}{2}\dot{\omega}_0 t^2$$

$$= -\omega_0\tau - \dot{\omega}_0\tau t + \dot{\omega}_0\tau^2 = -\tau\left(\omega_0 + \dot{\omega}_0(t - \tau)\right)$$

or

$$\Delta\Psi = -\tau\omega_1.$$

Therefore, in the transfer function of the delay $W_d(s) = e^{-\tau s}$, for the input given by (6.27), the frequency response of the delay can be obtained through the substitution of $j\omega_1$ for s, where ω_1 is the instantaneous frequency of the delayed signal. We note that the amplitude response is the unity.

Now we shall analyse the response of the delay to the decaying sinusoid

$$u(t) = e^{\sigma t} \cos(\omega t). \tag{6.28}$$

The delayed control is

$$u_1(t) = e^{\sigma(t-\tau)} \cos\big(\omega(t-\tau)\big).$$

While the phase shift introduced by the delay is, obviously, $\Delta\Psi = -\tau\omega$, the gain can be determined as the ratio of the amplitudes:

$$K = \frac{e^{\sigma(t-\tau)}}{e^{\sigma t}} = e^{-\sigma t}.$$

Therefore, in the transfer function of the delay $W_d(s) = e^{-\tau s}$, for the input given by (6.28), the frequency response of the delay can be obtained through the substitution of s with $j\omega$.

Combining the results obtained separately for the variable frequency input and for the decaying input, we can formulate the following rule for the computation of the frequency response of the delay: *In the transfer function of the delay $W_d(s) = e^{-\tau s}$, s must be replaced with $\sigma + j\omega_1$, where σ is the instantaneous decay and ω_1 is the instantaneous frequency of the delayed signal, which in turn is equal to $\omega(t) - \tau\frac{d\omega(t)}{dt}$.* One can see that the rate of frequency change does not affect the amplitude characteristic, and the decay does not affect the phase shift, so that a separate account of the decay and varying frequency that was undertaken above is possible.

Therefore, the DHB principle for the system with delay in the loop can be written as follows. The magnitude balance equation is

$$\big|N(a,\sigma)\big|e^{-\tau\sigma}\big|W^*(\sigma,\omega,\dot{\omega})\big| = 1. \tag{6.29}$$

The phase balance equation is

$$\arg N(a,\sigma) - \tau\omega + \tau^2\dot{\omega} + \arg W^*(\sigma,\omega,\dot{\omega}) = -\pi. \tag{6.30}$$

And the rates of change balance is given by

$$\frac{\partial \ln|W^*(\sigma,\omega,\dot{\omega})|}{\partial \ln\omega}\frac{\dot{\omega}}{\omega} = -\frac{\partial \ln N(a,\sigma)}{\partial \ln a}\sigma, \tag{6.31}$$

where $W^*(\sigma,\omega,\dot{\omega})$ is the modified frequency response of the linear part considered above (conventional frequency response with $s = \sigma + j\omega$ closed by the feedback containing $j\dot{\omega}$), which should not include the delay.

6.5 Describing Function of MRFT for Sinusoidal Input of Exponentially Changing Amplitude

We now derive the describing function of the MRFT for the sinusoid input of exponentially increasing (decreasing) amplitude. We use formulas (6.8) for computing

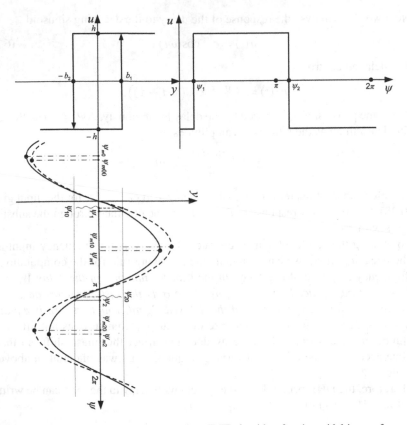

Fig. 6.1 Derivation of describing function for MRFT algorithm for sinusoidal input of exponentially changing amplitude

the describing function. The methodology of the derivation of the DF for a sinusoidal input of exponentially varying amplitude is similar to the derivation of the conventional DF. Some formulas for the DFs of second-order sliding mode control algorithms relevant to the current task were presented in publications [20] and [69], including the suboptimal algorithm in the latter.

Consider the following diagram, which is useful to the derivation of the DF formula (Fig. 6.1).

It is worth noting that we will be able to use the formula for the conventional DF of the MRFT obtained earlier if we account for the effect of the delay between the time of setting of the value of the hysteresis b_1 or b_2 (time t_{mi} and corresponding phase Ψ_{mi}) and the time of the switch that uses this value of the hysteresis (time t_i and corresponding phase Ψ_i). We introduce the variable which we shall name *true beta* β_{tr}. True beta gives the ratio between the value of the hysteresis at the time of the switch to the value of the instantaneous amplitude at the same very time:

$$\beta_{tr} = \frac{b(t_i)}{a(t_i)}.$$

Considering now the exponential character of the amplitude change and assuming $a(t) = a_0 e^{\sigma t}$ write for time instants t_{mi} and t_i:

$$a(t_{mi}) = a_0 e^{\sigma t_{mi}},$$

$$a(t_i) = a_0 e^{\sigma t_i}.$$

The relationship between b and $y_{mi} = a(t_{mi})$ established through the MTFT algorithm is given by $b(t_i) = \beta a(t_{mi})$. Therefore,

$$\beta_{tr} = \frac{b(t_i)}{a(t_i)} = \frac{\beta a(t_{mi})}{a(t_i)} = \beta \frac{a_0 e^{\sigma t_{mi}}}{a_0 e^{\sigma t_i}} = \beta e^{\sigma(t_{mi}-t_i)}.$$

Express the time shift through the phase shift and the frequency of the signal:

$$t_{mi} - t_i = \frac{\Psi_{mi} - \Psi_i}{\omega}.$$

Now taking into account the fact that for the periodic motion $\Psi_i = \arcsin\beta$, and $\Psi_{mi} = -\frac{\pi}{2}$ (see Fig. 6.1) we can obtain the following expression for *true beta*:

$$\beta_{tr} = \beta e^{-\sigma(\frac{\pi}{2}+\arcsin\beta)/\omega}.$$

Substitution of the true beta expression for β in the DF formula of MRFT (3.3) yields

$$N(a, \sigma, \omega) = \frac{4h}{\pi a}\left[\sqrt{1 - \beta^2 e^{-2\sigma(\frac{\pi}{2}+\arcsin\beta)/\omega}} - j\beta e^{-\sigma(\frac{\pi}{2}+\arcsin\beta)/\omega}\right]. \quad (6.32)$$

Formula (6.32) is the describing function for the MRFT. Interestingly enough, the magnitude of this function depends only on the amplitude:

$$\left|N(a, \sigma, \omega)\right| = \left|N(a)\right| = \frac{4h}{\pi a}, \quad (6.33)$$

and the phase characteristic depends on the σ to ω ratio and does not depend on the amplitude:

$$\arg N(a, \sigma, \omega) = \arg N(\sigma/\omega) = -\arctan \frac{\beta e^{-\sigma(\frac{\pi}{2}+\arcsin\beta)/\omega}}{\sqrt{1 - \beta^2 e^{-2\sigma(\frac{\pi}{2}+\arcsin\beta)/\omega}}}$$

$$= -\arcsin\left(\beta e^{-\sigma(\frac{\pi}{2}+\arcsin\beta)/\omega}\right). \quad (6.34)$$

One can see that setting the value of σ to zero results in the conventional DF formula for the MRFT obtained above.

6.6 Dynamic Harmonic Balance in System with MRFT Algorithm

We found in the previous section that the DF of MRFT is a complex quantity. At the same time we obtained the conditions of DHB in the system with one single-valued nonlinearity. We now extend the DHB application to the system with MRFT algorithm.

In the beginning of this chapter we considered the controller to be represented by a single-valued symmetric nonlinearity; that equation (6.17) providing the relationship between the amplitude decays in the input and output signals of the nonlinearity was derived for N being a real quantity. In the case of the MRFT, the describing function is a complex quantity, and equation (6.17) needs to be modified as follows.

$$\sigma_u = \frac{\dot{a}_u}{a_u} = \frac{\frac{\partial |N|}{\partial a} \dot{a} a + |N| \dot{a}}{|N| a} = \frac{\partial \ln |N|}{\partial a} \dot{a} + \sigma = \left(\frac{\partial \ln |N|}{\partial \ln a} + 1 \right) \sigma. \qquad (6.35)$$

Considering (6.33), we can see that

$$\frac{\partial \ln |N|}{\partial \ln a} = -1$$

and $\sigma_u = 0$, which totally agrees with the fact that the output signal of the MRFT (control) has constant amplitude of oscillations.

With formulas derived for the DF of MRFT, we can write the DHB equations for MRFT as follows. The magnitude balance equation is

$$\frac{4h}{\pi a} e^{-\tau \sigma} \left| W^*(\sigma, \omega, \dot{\omega}) \right| = 1; \qquad (6.36)$$

the phase balance equation is

$$- \arcsin\!\left(\beta e^{-\sigma(\frac{\pi}{2} + \arcsin \beta)/\omega} \right) - \tau \omega + \tau^2 \dot{\omega} + \arg W^*(\sigma, \omega, \dot{\omega}) = -\pi; \qquad (6.37)$$

and the rates of change balance is

$$\frac{\partial \ln |W^*(\sigma, \omega, \dot{\omega})|}{\partial \ln \omega} \frac{\dot{\omega}}{\omega} = \sigma, \qquad (6.38)$$

where $W^*(\sigma, \omega, \dot{\omega})$ is the modified frequency response of the linear part considered above (conventional frequency response with $s = \sigma + j\omega$ closed by the feedback containing $j\dot{\omega}$), which should not include the delay.

6.7 Example of Analysis of Transient Motions Through Dynamic Harmonic Balance

Consider a rather challenging example of application of the dynamic harmonic balance principle to analysis of the transient oscillations in a system, the block diagram of which is presented in Fig. 6.2, with parameter values: $k_1 = 9.76$, $k_2 = 0.981$, $\rho = 1$, and initial value of the state vector $x = 0.05$, $\dot{x} = 0$. The complexity of the dynamics of the system being considered is related to the fact that oscillations generated in the system have zero or very low frequency at the beginning and approach infinite frequency at the end. For the amplitude, vice versa: it starts from some finite value and approaches zero with time approaching infinity. The presented model describes the dynamics of a mechanical system that was analysed in [28]. The nonlinearities of the system are $f_1(x) = \sin x$ and $f_2(x) = \text{sign}(x) \cos(x)$. The second nonlinearity is shown in Fig. 6.3.

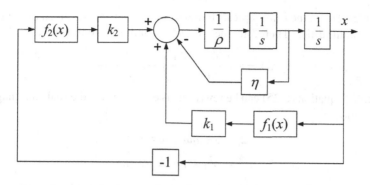

Fig. 6.2 Dynamic model of system

Fig. 6.3 Second nonlinearity

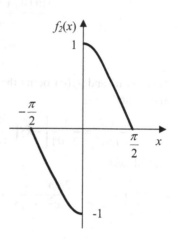

The describing functions for these two nonlinear functions are obtained as follows [8, 28]:

$$N_1(a) = \frac{2J_1(a)}{a},$$

where $J_1(a)$ is the Bessel function of the first order, and

$$N_2(a) \approx \frac{4}{\pi a} - \frac{8a}{3\pi}.$$

We can now carry out analysis of the motions in the system using the DHB principle. The transfer function of the linear part is given as follows:

$$W_l(s) = \frac{1}{(\rho s + \eta)s}.$$

The modified frequency response can be written for $W_l(s)$ as

$$W_l^*(\sigma, \omega, \dot{\omega}) = \frac{1}{\rho(\sigma + j\omega)^2 + \eta(\sigma + j\omega) + j\dot{\omega}}.$$

The nonlinearity in the Lure representation of the system can be given as the sum of the two nonlinearities. Therefore, the DF of the combined nonlinearity is

$$N(a) = k_2 \left(\frac{4}{\pi a} - \frac{8a}{3\pi} \right) - k_1 \frac{2J_1(a)}{a}.$$

Complex equation (6.25) can be rewritten as equations for the real and imaginary parts as follows:

$$2\rho\sigma\omega + \eta\omega + \dot{\omega} = 0, \tag{6.39}$$

$$\rho(\sigma^2 - \omega^2) + \eta\sigma = -N(a). \tag{6.40}$$

And we obtain the third algebraic equation per (6.26), considering that

$$\frac{\partial \ln|W_l^*(\sigma, \omega, \dot{\omega})|}{\partial \omega} = \frac{2\omega\rho}{\rho(\sigma^2 - \omega^2) + \eta\sigma},$$

$$\frac{dJ_1(a)}{da} = \frac{J_0(a) - J_2(a)}{2},$$

with $J_0(a)$ and $J_2(a)$ being the Bessel functions of zero and second order, respectively, and

$$\frac{\partial \ln N}{\partial \ln a} = \frac{1}{N(a)} \left[-k_2 \left(\frac{4}{\pi a} + \frac{8a}{3\pi} \right) + k_1 \left(\frac{2J_1(a)}{a} - J_0(a) + J_2(a) \right) \right],$$

as follows:

$$\sigma - \frac{2\omega\rho}{\rho(\sigma^2 - \omega^2) + \eta\sigma} \dot{\omega} = \left[\frac{1}{N(a)} \left(-k_2 \left(\frac{4}{\pi a} + \frac{8a}{3\pi} \right) \right. \right.$$
$$\left. \left. + k_1 \left(\frac{2J_1(a)}{a} - J_0(a) + J_2(a) \right) \right) + 1 \right] \sigma. \tag{6.41}$$

Expressing $\dot{\omega}$ from (6.39) and substituting into (6.41), we obtain two equations with three unknown variables. We also denote

$$g_1(a) = -N(a) = -k_2 \left(\frac{4}{\pi a} - \frac{8a}{3\pi} \right) + k_1 \frac{2J_1(a)}{a}$$

and

$$g_2(a) = \frac{1}{N(a)} \left[-k_2 \left(\frac{4}{\pi a} + \frac{8a}{3\pi} \right) + k_1 \left(\frac{2J_1(a)}{a} - J_0(a) + J_2(a) \right) \right] + 1$$

$$= \frac{1}{N(a)} k_1 (J_2(a) - J_0(a)).$$

With this notation the set of the two equations can be written as

$$\rho(\sigma^2 - \omega^2) + \eta\sigma = g_1(a),$$

$$\sigma + \frac{2\omega^2 \rho(2\rho\sigma + \eta)}{\rho(\sigma^2 - \omega^2) + \eta\sigma} = g_2(a)\sigma. \tag{6.42}$$

Fig. 6.4 Transient motions in the system

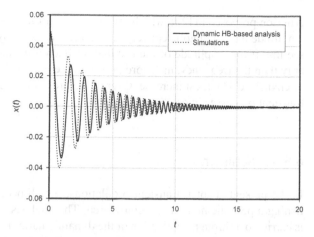

We shall solve (6.42) as a set of equations for σ and ω, considering amplitude a a given parameter. With this approach, if we express ω from the first equation of (6.42) and substitute in the second equation, we arrive at one algebraic equation with one unknown variable σ:

$$\frac{2\rho(2\rho\sigma + \eta)(\sigma^2 - \frac{g_1(a)-\eta\sigma}{\rho})}{\sigma g_1(a)} = g_2(a) - 1,$$

which can be transformed into a third-order algebraic equation for σ:

$$c_0 + c_1\sigma + c_2\sigma^2 + c_3\sigma^3 = 0, \tag{6.43}$$

where $c_0 = -2g_1(a)\eta$, $c_1 = -4\rho g_1(a) + 2\eta^2 - g_1(a)(g_2(a) - 1)$, $c_2 = 6\rho\eta$ and $c_3 = 4\rho^2$.

Solution of the cubic equation (6.43) is well known. In the considered example, we can use the Cardano's formula, which allows one to analytically find the real root of equation (6.43). This is possible due to the fact that σ is a real quantity. With a and σ available, we can numerically integrate the differential equation (6.10). After doing so ω is computed through the formula that was used for the substitution (based on (6.40)). Equations (6.39)–(6.41) are thus solved at every step of the integration of (6.10), and the time response of the system is obtained.

The initial amplitude $a(0)$ can be accepted equal to $x(0)$ because $\dot{x}(0) = 0$ and the former is the point of maximum of $x(t)$. The transient process in the system is shown in Fig. 6.4 for the presented DHB-based solution and simulations based on the original differential equations.

One can see that the presented approach provides a good estimate of the transient dynamics even for this challenging dynamic system. In fact there is only some phase mismatch accumulated at the beginning of the transient. Later this phase shift does not change.

What distinquishes the DHB-based approach from the solution of the original differential equations is the explanation of the mechanism of the evolution of the parameters of the oscillations and the possibility of the analysis of asymptotic behaviour of the system. It can be found in particular that, in accordance with the criterion proposed in [26], which involves the notion of the *phase deficit*, oscillation amplitude has asymptotic convergence to zero in the analysed system. One can also see that if the amplitude of the oscillations decreases due to the dissipation of energy then the frequency, in accordance with the equation of the balance of the rates of change (6.41), must increase. Overall the presented model conveniently explains transient processes in oscillatory systems.

6.8 Conclusions

In the present chapter, transient oscillations, which occur in the process of establishing a periodic motion, are considered. The analysis of the transient oscillations is carried out through application of the dynamic harmonic balance principle, which is also presented in the chapter. The dynamic harmonic balance is treated as an extension of the conventional harmonic balance, which is applied to a periodic motion, to the analysis of an oscillatory motion of variable frequency and amplitude. The equations of the magnitude balance and phase balance are modified and completed with an equation of balance of rates of change of amplitude and frequency. The dynamic harmonic balance for process models containing delay is also considered. Derivation of the describing function for the exponentially changing amplitude for the modified relay feedback test is presented, and the dynamic harmonic balance for the system with MRFT algorithm is formulated. Application of the dynamic HB principle is illustrated by an example of analysis of transient motions in a second-order nonlinear dynamic system.

Chapter 7
Software for Loop Tuning in Distributed Control Systems (DCS)

Practical aspects of the implementation of the modified relay feedback test and other algorithms in industrial loop tuning software are discussed. Such problems as the dependence of test software residence on performance, and the existence and the effects of noise, disturbances and process nonlinearities are analysed. An example of industrial loop tuning software is given.

7.1 Specifics of Loop Tuning in DCS

Processes at modern industrial plants are usually controlled by distributed control systems (DCS). DCS have a distributed computing power architecture, where every part of the process is controlled by a separate controller and all the controllers are connected with each other (both "directly" through *peer-to-peer* connections and "indirectly" through higher level controls) and with servers and stations. Normally the loop must be closed within one controller, which ensures high reliability and faster computer processing. In rare situations, a loop may be closed through the use of two or more controllers. In the latter case peer-to-peer connections or redundant networks are used to ensure reliable data exchange between these controllers. A typical DCS architecture is presented in Fig. 7.1. Servers and operator/engineering stations equipped with human-machine interface (HMI) displays are an integral part of a DCS, which provides the possibility of monitoring and controlling the process from operator stations as well as of programming control algorithms and loading these algorithms into the controllers. Algorithms loading into controllers can usually be done without shutting down the process, through retentive memory that ensures "freezing" the controller outputs for the time of the algorithm load.

Loop tuning software may be run on either a controller or a server (see Fig. 7.1). In the latter case this server is usually a specialised one, which is sometimes called the *applications server*. If the loop tuning software resides on a server then the mechanism of accessing the respective I/O by the tuning software with the purpose of manipulating the output signal is usually indirect: the respective PID controller

I. Boiko, *Non-parametric Tuning of PID Controllers*, Advances in Industrial Control, DOI 10.1007/978-1-4471-4465-6_7, © Springer-Verlag London 2013

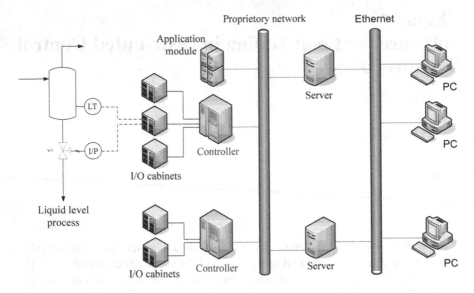

Fig. 7.1 Typical architecture of DCS

module[1] is switched into "Manual" mode and the controller output is manipulated remotely by the loop tuning software. After the process of tuning is complete or interrupted, the PID controller is switched back to its normal mode ("Automatic" or "Cascade"). Upon completion of tuning and obtaining the PID controller module tuning parameters, they can be either manually entered into the controller settings or transferred to the PID controller module by the tuning software.

If the tuning software resides in the controller then it is usually integrated into the PID controller module software. We can see this approach in the Emerson *DeltaV*® DCS. Alternatively it can be programmed in the same way as control algorithms. However, DCS software is hardly suitable for conventional computer programming (loops, logical conditions, iterative calculations, etc.). In the author's experience, the respective pieces of software realised by means of DCS programming software appear bulky, become more complex than the process control software itself, can usually be applied to only one PID controller module (another one would require a similar piece of software) and can only be suitable if some PID controller module definitely needs tuning or auto-tuning functionality.

Server-residing software is usually implemented as a tuner, whereas controller-residing software can also be implemented as auto-tuner. The difference between the two is only in the method of tuning initiation and setting of PID controller module parameters. In loop tuners, test initiation is done manually, and the results of tuning can either be entered manually or transferred upon issue of respective command.

---

[1]To differentiate between a physical controller and a PID controller (configurable software block or module) in a DCS, we shall refer to the latter as to a PID controller module.

In autotuners, tuning can be programmed to be executed in a periodic manner, and results of tuning are automatically transferred to the PID controller module and immediately used by it.

The main advantage of the controller-residing software is a higher performance of the tuning software due to high sampling rates of the I/O cards and low execution period of the controller itself. With modern DCS hardware such rates as 100 ms of the I/O sampling rate and a 100-ms execution period of the controller are easily achievable, which introduces a very small delay into the tuning algorithm. With server-residing tuning software, performance is limited mostly by the access time of the controller variables by the server software, which may reach a few seconds. It may be critical for tuning such fast loops as certain flow and pressure loops. These circumstances need to be considered in tuning software design.

7.2 Methods of Mitigating Effects Existing in Real Processes

7.2.1 Invasive Character of Tuning

Most loop tuning methods involve some type of test on the process: step test, RFT, etc. Because tuning is usually done on a live process during normal plant operation, tuning has an invasive character and creates certain upsets to the process, which may propagate to other processes, too. There some noninvasive methods of tuning, which do not involve any tests but use process observations instead. They are normally referred to as *noninvasive*. They use various events resulting in process disturbances for process response measurement and subsequent loop tuning.

The MRFT and other tests considered in this book, of course, create a disturbance of the process, which is an undesirable but necessary feature. Referring to MRFT we note that the value of this disturbance is proportional to the relay amplitude. From this point of view, smaller amplitudes of the relay should be preferred. The choice of the amplitude is limited from below by considerations of noise and the presence of external disturbances. The MRFT algorithm must be able to excite oscillations in the process loop even in the presence of noise and disturbances. And the oscillations excited by the MRFT must be "visible" in the background containing noise and disturbances. Therefore, the choice of amplitude has an effect on the *resolution* of the method, which becomes higher if the relay amplitude is larger.

Another circumstance is the *controllability*[2] of the system during the test. To illustrate this let us consider tuning with the use of the step test. Because the step test is applied to the process the loop is opened for the duration of the test, and the process is not controlled to the set point during that time. If a certain event occurs at the time of tuning, which results in a process disturbance, then the disturbance

---

[2]The term *controllability* is used here not in the mathematical sense but as the property of whether the process variable is controlled to a specified set point during the test.

effect would not be attenuated (as in the closed loop situation). This problem is solved to some extent in the RFT/MRFT. When the process is controlled by MRFT the loop remains closed and the process variable (controlled variable) is controlled in the discontinuous manner by the bang-bang-type controller to the set point. If a disturbance occurs at the time of test, the disturbance would be attenuated by this control: the control instead of being symmetric will become asymmetric, having different duration of positive and negative pulses. Yet, the process variable is still controlled. This is a definite advantage of all continuous cycling methods over the methods that use the step test. However, if the disturbance is so high that it disrupts the oscillations in the loop, then the MRFT algorithm produces a constant output h or $-h$ and the loop becomes open. It is worth noting that the higher the value of the amplitude h, the smaller the probability for this event to happen. This is related to the range of average control values that a controller with particular amplitude is capable of producing. This range is proportional to amplitude h.

It becomes obvious that there is a trade-off between the desire to have a small amplitude of the MRFT test so as not to disturb the process and the danger of the situation of the loop opening caused by a sufficiently high disturbance, thus losing process controllability. Some special measures were found helpful to prevent the situation of loss of control to set point. This is especially relevant to auto-tuners, where the tuning process is not monitored. Setting maximum waiting time for the oscillation (or next oscillation/relay switching) to occur and the maximum time for the whole tuning would be useful in this respect. This waiting time is different, of course, for different processes, and some adjustments of this parameter or parameters are necessary. If the event of loop opening during tuning is detected then the software must ensure switching to normal PID control with PID parameters unchanged.

7.2.2 Mitigation of Noise Effects

Noise mainly in the form of measurement noise is present in every control system. It is caused by the inherent features of the selected method of measurement of a particular process variable and interference with electromagnetic fields created by various equipment. The effect of the second source of noise is usually reduced by limiting the length of instrumentation wires, shielding and laying of instrumentation wiring farther away from power wiring, which is the source of the electromagnetically induced noise.

Noise created by the method of measurement itself is usually difficult to reduce. For example, if flow is measured through the differential pressure across an orifice plate, the orifice plate creates local turbulence of the flow, which is revealed as differential pressure fluctuations and received by the DCS as noise. Changing the method of measurement to, for example, vortex or Coriolis might reduce the level of noise but be a more expensive option and, therefore, less preferable. The transmitter noise can easily be reduced in the transmitter itself or in a DCS through low-pass filtering. However, the introduction of a low-pass filter in the loop brings additional undesirable lag into the loop, which worsens the loop dynamics. Because of this, the degree of noise reduction through low-pass filtering is very limited. And, if filtering is used in the loop, it is important that the same filter should be used when the MRFT is run.

The same noise that affects the PID loop performance affects the functionality and accuracy of MRFT and subsequent measurements and tuning. First, the presence of noise may cause false switchings of the relay. And second, the presence of noise results in the switching of the relay not by the true (averaged in a certain sense) process variable signal but either a slightly higher or slightly lower value.

The first effect is usually mitigated by the use of hysteresis on the relay characteristic. We can find this solution in [40] and many other references on relay control. This is a known industry-recognised solution. However, one can see that the introduction of the hysteresis in the conventional RFT results in a change of the frequency and the amplitude of oscillations (the value of $-\frac{\pi b}{4h}$ provides a different point on the LPRS than the real axis). Small hysteresis values may possibly not distort measurements too much, but if the measurement signal is noisy then the required value of the hysteresis would be high enough to result in a high enough error of the frequency and amplitude measurement.[3]

The use of MRFT normally involves some hysteresis in the relay characteristic. However, first, this hysteresis is a calculated value and can be small—as we can see from the optimal tuning rules produced for the level process. Second, if a proportional controller is used then the hysteresis must be zero. And third, if a PID controller is tuned then the hysteresis value in MRFT is small. As the industrial loop tuner design work showed, noise protection logic is a necessary feature of loop tuning software.

Control logics equivalent to the ideal relay were developed to provide noise-insensitive functionality. The logics (see Fig. 7.2) includes two hysteretic relays with adjustable hysteresis values and hysteresis shifted in the positive direction and in the negative direction (*Relay 1* and *Relay 2*), respectively; two analog-to-Boolean value converters *A/D Converter*, two short pulse generators *PG*; *RS-trigger*; and a signal selector (*Selector*) controlled by the SR-trigger, which selects either the output of the first relay or the output of the second relay. The logics are designed in such a way that when the error signal $\sigma(t)$ grows the second relay is activated (selected), and when $\sigma(t)$ decreases the first relay is active. Switching is thus realised in accordance with the characteristic of the ideal relay. In the MRFT algorithm, which involves some hysteresis in the switching, the characteristics of Relay 1 and Relay 2 are shifted right or left from those shown in Fig. 7.2 to ensure the necessary hysteresis value of the algorithm itself.

Another method of protection from noise is the use of switching inhibition through the introduction of a delay. The logic behind it is that every subsequent switch is inhibited during a certain period of time after the previous switch.

---

[3]The frequency and amplitude will in fact be measured at a different point of the Nyquist plot or LPRS.

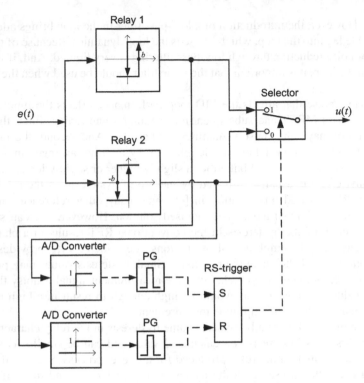

Fig. 7.2 Equivalent ideal relay characteristic with noise protection

This time delay can be either hard-coded or be a user-defined parameter. Both methods of noise protection are used in the loop tuning software considered below.

Both presented methods can protect the MRFT from false switches or chattering caused by noise. However, the problem of the relay switching by the signal that is not a true process variable still remains unsolved. This effect can, however, be mitigated by using a few cycles for measurement and averaging the results. Indeed, noise comes to the DCS as a random value with some statistics, among which we can legitimately assume the average being zero. If the average is nonzero it is then a systematic error of the process variable measurement and should be handled in a different way. The chance that the relay is switched by a slightly higher than the true value is the same as the chance that the relay is switched by the signal that contains the noise component of the same magnitude but negative sign. When we consider a few periods of the test oscillations we can average the effect of noise and make the generated frequency insensitive to noise in the average sense. The noise still affects the amplitude of the signal, and this problem can only be mitigated by the use of sufficiently high amplitude of the relay or the use of low-pass filtering of the error signal, and by accounting for the filter lag in the calculations of β and the amplitude.

7.2.3 *Mitigation of Effect of External Disturbances*

Disturbances are inevitable in every control system. In process control systems, disturbances are the main reason for automation because most control loops work not in the servo mode but in the regulator mode. Disturbances come from other parts of the plant or process in the form of variable feed rates or consumption rates, from the environment (e.g., varying temperatures) or generated within the unit itself by switching on and off pumps, and other equipment, throttling lines connected to the line associated with the process under control, etc. We can see the mechanism of disturbance generation from the analysis of processes considered in Chap. 2.

Generally we consider disturbances to be constant or slowly varying (slowly in comparison with the transients in the PID loop or MRFT test). If the disturbance is fast-changing then its effect may be considered the same as the effect of noise. The effects of fast-changing disturbances can also be mitigated similarly to the those of noise. In constant disturbances, though, we can find the initial time interval right after the occurrence of the event producing it, and the subsequent time. For example, after starting the second pump in the two-pump arrangement the flow through the line my quickly grow due to the change of the combined pump performance curve, so that the disturbance comes to the flow loop as the increment of the differential pressure across the control valve. An important observation for the flow loop that can be drawn from this example is that the disturbance is not the differential pressure itself but the increment of the differential pressure. However, the increment is defined by the operating point and, therefore, once the loop comes to a new steady state the definition of the operating point can be revised so that the new condition can be considered an operating point with the respective value of the differential pressure across the valve. And the disturbance can be considered zero in this new operating point. Any new change of the differential pressure across the valve would thus produce a disturbance.

Effects of any disturbance produced during the test would be difficult to mitigate if the disturbance is significant. It may even disrupt oscillations, and the best option would probably be stopping the test and starting a new one after the process has stabilised at a new operating point. However, because in process industries the MRFT over the process can only be implemented in an incremental way, the disturbance is compensated for by the conditions at the operating point. As we saw in the example of the flow loop, in every steady operating condition the disturbance may be considered zero, and every change from this condition may be considered a disturbance. These circumstances prompt a method of treating disturbances in MRFT that involves initially bringing the process controlled by a PID controller (not tuned yet or at least not optimally tuned) to a steady state. After which the MRFT can be applied to the process via controller output changes from the steady state by $\pm h$. If the external disturbance is constant, its effect would, therefore, be compensated for by the correct initialisation before the test.

Analysis of the effect of external disturbances of the MRFT can be done through the use of the LPRS method presented in Chap. 5.

7.2.4 Accounting for Process Nonlinearities

We are going to consider here only static (single-valued) nonlinearities found at the process input and output or those that can be transposed to the process input or output. Both reveal themselves as nonlinear process gain or the process gain that depends on the choice of the operating point. Output nonlinearities can be easily linearised through applying a proper nonlinear function to the sensor (transmitter) signal. An example of this approach is the square-root linearisation of the differential pressure signal for flow measurement. An example of the input nonlinearity is the nonlinear characteristic of the valve, which can also be linearised in the controller through using an inverse nonlinearity. This, however, is only an approximate treatment, which is valid if the valve dynamics are neglected. Physically, this non-equivalence is related to a difference in valve travel at the original control and the control propagated through the linearising nonlinearity. Yet, if the valve is much faster than the process then this linearisation is acceptable.

Both considered types of nonlinearities are always present in any process control system. Sensors do not provide ideal linearity, and the so-called installed characteristics of valves are always nonlinear even if the theoretical characteristic is linear. Also, even if all the possible linearisations are done, there are still uncompensated or residual nonlinearities that need to be considered. The effect of static nonlinearities in the process on the MRFT is seen as asymmetric (unequally spaced) oscillations of the control signal. This is a result of the "effective" (averaged) process gain being different for positive and negative control pulses. In that respect the effect of a nonlinearity is similar to the effect of an external disturbance, which also introduces nonzero average into the control.

It is easy to show that the presence of a static nonlinearity at the input of the process does not affect the frequency of the oscillations in an MRFT (see the detailed research on this in [15]). Yet the amplitude of the oscillations becomes affected. For this reason, half the double amplitude $a = 0.5(a_p + |a_n|)$ must be used in calculations of PID controller tuning parameters (see Chap. 5 for details).

7.3 DCS Loop Tuning Software Description

Software *CLTune* for tuning loops in DCS Honeywell *TPS*® was developed on the basis of the MRFT. Two other tests can be used in this software, too, but we give the description pertaining only to the MRFT. The MRFT algorithm resides in the controller (high performance process manager or HPM) and the HMI part is installed on a server. Through this arrangement both speed of computing and convenience of use are ensured. The HMI part must be installed on every server or on the station where the software is going to be used.

There are a number of settings used in the algorithm of CLTune. Parameter selection is done via "pressing" buttons on the HMI pages. There are two buttons: "Settings" and "Trend" available on Settings and Trend pages, which are used for

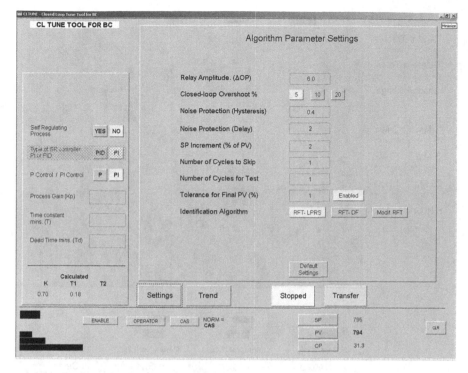

Fig. 7.3 Settings page of software CLTune. Courtesy of Syncrude Canada

switching between those pages (see Fig. 7.3). The following selections are available on the Settings page.

- Self-regulated and non-self-regulated process. For temperature, flow and analyser loops, the self-regulated type of process should be selected; for level loops, the non-self-regulated type of process should normally be selected; pressure loops can include either self-regulated or non-self-regulated type of process.
- For self-regulated type of the process, two types of controllers can be tuned: proportional-integral (PI) and proportional-integral-derivative (PID).
- For non-self-regulated type of the process, two types of controllers can be tuned: proportional (P) and proportional-integral (PI).

Selectable parameters that are used for defining the test-related values are as follows.

- Tuning Algorithm. This is one of the following: Asymmetric RFT (see [17]), RFT or MRFT.
- Relay Amplitude is the parameter that defines an increment of the controller OP [%] in the test, with respect to the value immediately before the test.
- Required loop response can be selected as Slow, Moderate or Fast by pressing corresponding buttons. For MRFT tuning method, they are quantified in terms of the required gain stability margin as 4, 3 and 2 respectively.

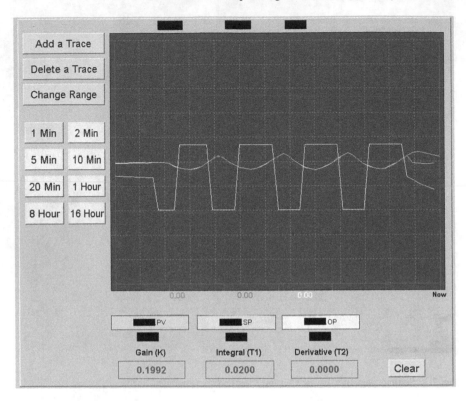

Fig. 7.4 Trends page of software CLTune. Courtesy of Syncrude Canada

- Noise Protection (hysteresis) is the hysteresis value in the relay characteristic introduced in the control loop (see Sect. 7.2.2 about noise protection).
- Noise Protection (delay) is the value of time interval after the last switch of the relay within which inhibiting of the next switch of the relay is activated. It can be entered in seconds in the respective field.
- SP Increment does not apply to MRFT.
- Number of Cycles to Skip is the number of periods of the oscillations that will be ignored before the actual measurements of the parameters of those oscillations begin. This is aimed at allowing for the oscillatory process to be established, so that the transient part of the process can be ignored. At least one cycle must be skipped in the test.
- Number of Cycles for Test is the number of periods of the oscillations that will be used for measurements and subsequent calculations of tuning parameters.
- Tolerance for Final PV does not apply to MRFT.
- Button Enabled is not used in MRFT tuning.
- Button Default Settings allows for setting all above-referenced parameters to default values.
- Button Transfer is used to transfer the calculated tuning settings into the respective PID controller.

The Trends page (see a fragment of the page in Fig. 7.4) is intended for displaying in real time the process variable, set point and controller output of the selected loop. Other variables, even those not pertaining to the selected loop, can be chosen for display, too. The time scale and the displayed range for each variable can be set up.

Test and tuning is initiated by pressing the button "Stopped", which immediately after starting the test changes its value to "Run", indicating that the test is running. Upon completion, the calculated tuning variables are displayed on both Settings and Trend pages. They can be transferred to the controller setting by pressing the button Transfer.

A sample trend of a liquid flow loop tuning is provided in Fig. 7.4.

7.4 Conclusions

In this chapter, aspects of the implementation of the MRFT-based loop tuning method in distributed control systems are considered. Such circumstances as noisy signals, the presence of disturbances and nonlinearities, possible consequences of incorrect selection of the relay amplitude and consequences of the occurrence of an event that results in a large disturbance application at the time of running the test are analysed. These aspects of loop tuner design are crucially important to creation of high-performance loop tuning software.

Possible residence of the software in the DCS is discussed, with analysis of advantages and drawbacks of each option. And, finally, an industrial loop tuner based on the MRFT is presented. Its functionality and main features are described. The provided material may be very useful to those involved in loop tuning software design.

Chapter 8
Appendix

8.1 Sample Simulink Models Used in Optimisation

Simulink models used with the MATLAB code for finding parameters of the optimal tuning rules are presented in Figs. 8.1, 8.2 and 8.3.

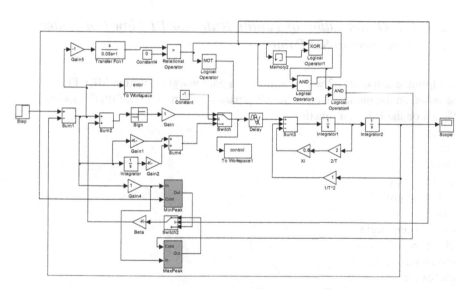

Fig. 8.1 Model of flow loop used in optimisation of PI tuning rules

I. Boiko, *Non-parametric Tuning of PID Controllers*, Advances in Industrial Control,
DOI 10.1007/978-1-4471-4465-6_8, © Springer-Verlag London 2013

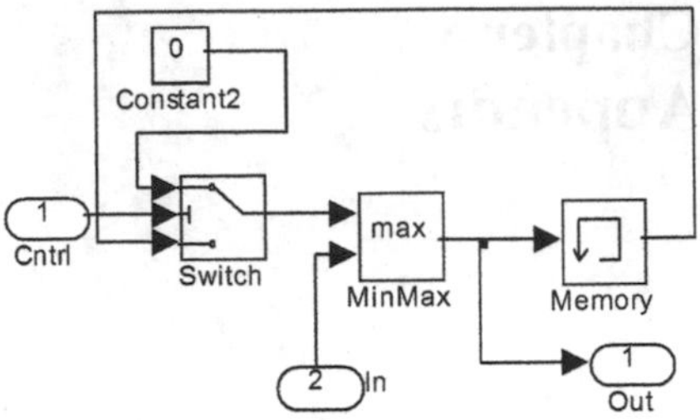

Fig. 8.2 Model of max peak
detector used in MRFT
Fig. 8.1

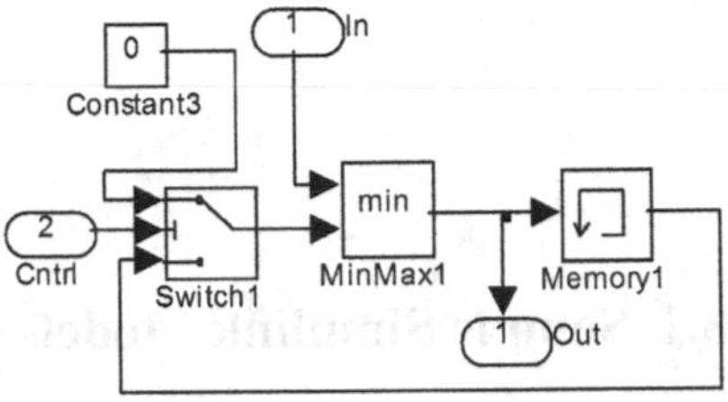

Fig. 8.3 Model of min peak
detector used in MRFT
Fig. 8.1

8.2 Matlab Code Used in Book

8.2.1 ISE Optimisation of Tuning Rules for PI Flow Controller (Response to Set Point)

```
% Optimisation of c1,c2 with constraint on gain margin (through MRFT)
% based on the model given in MRFT_2param_optim.mdl (DSOPDT model)
% and on the ISE cost function
script
clc
clear
global q_opt
global control
global error
global gainmargin
global beta
global omega_u
global tout
% Selecting required gain margin
gainmargin=2.;
% Selecting dead time and damping of the model
deadtime=1.5; % tau of DSOPDT model
xi=0.8; % damping of DSOPDT model
% Opening Simulink model
open_system('MRFT_2param_optim')
% Setting dead time
set_param('MRFT_2param_optim/Delay','Delay',num2str(deadtime))
```

```
% Setting xi (damping)
set_param('MRFT_2param_optim/Xi','Gain',num2str(xi));
X=fminbnd('mrft_opt_ise',0.1,0.9); % ISE optimisation
coefficient2=X %c2
coefficient1=1./(gainmargin*sqrt(1+1./(4*pi*pi*coefficient2^2))) %c1
q=q_opt
beta_opt=beta
% Computing equivalent time constant T_e of FOPDT model
t_e=tan(pi-atan(beta/sqrt(1-beta*beta))-deadtime*omega_u)/omega_u;
tau_te=deadtime/t_e

function q=mrft_opt_ise(X)
% ISE cost function
%
global q_opt
global control
global error
global gainmargin
global beta
global omega_u
global tout
% Setting c2
coefficient2=X;
% Computing c1
coefficient1=1./(gainmargin*sqrt(1+1./(4*pi*pi*coefficient2^2)));
% Computing beta
bet1=2*pi*coefficient2;
beta=1/bet1/sqrt(1-1/bet1/bet1);
% Selecting integration time step
timestep=0.05;
%
%%%%%%%%%%%%%%%%%%%%%%%%%%%%%%%%%%%%%%%%% %%%%%
% STEP 1 - RUNNING MTRFT
% Switching model to relay feedback test
set_param('MRFT_2param_optim/Constant','Value','1')
% Setting input to 0
stepinput=0.;
set_param('MRFT_2param_optim/Step','After',num2str(stepinput))
% Selecting and setting simulation time
simtime=100.0;
set_param('MRFT_2param_optim','StopTime',num2str(simtime))
% Setting beta
set_param('MRFT_2param_optim/Beta','Gain',num2str(beta));
% Run modified relay feedback test simulations
sim MRFT_2param_optim
```

```
% Finding switchings of control
npoints=simtime/timestep;
control0=control(11);
k=0;
for i=12:npoints
if control(i) =control0
k=k+1;
switching(k)=i;
control0=-control0;
end
end
% Run simulations again a longer time if the number of switches <9
if k<9
simtime=2*simtime;
set_param('MRFT_2param_optim','StopTime',num2str(simtime))
sim MRFT_2param_optim
npoints=simtime/timestep;
control0=control(11);
k=0;
for i=12:npoints
if control(i) =control0
k=k+1;
switching(k)=i;
control0=-control0;
end
end
end
% Ultimate period calculation
periodi=(switching(9)-switching(5))/2.;
ultperiod=periodi*timestep;
omega_u=2.*pi/ultperiod;
% Amplitude computing
maxerror=-1000;
for i=switching(7):switching(9)
if error(i)>maxerror
maxerror=error(i);
end
end
amplitude=maxerror;
% Ultimate gain computing
ultgain=4./pi/amplitude;
% PI controller parameters computing
propgain=coefficient1*ultgain;
inttimeconstant=coefficient2*ultperiod;
intgain=propgain/inttimeconstant;
```

```
%%%%%%%%%%%%%%%%%%%%%%%%%%%%%%%%%%%%%% %%%%%
% STEP 2 - SIMULATION OF STEP RESPONSE
% Switching the model to PID control
set_param('MRFT_2param_optim/Constant','Value','-1');
% Setting step value for step response
stepinput=1;
set_param('MRFT_2param_optim/Step','After',num2str(stepinput));
% Setting proportional and integral gains of PID controller
set_param('MRFT_2param_optim/Gain1','Gain',num2str(propgain));
set_param('MRFT_2param_optim/Gain2','Gain',num2str(intgain));
% Setting simulation time for step response
simtime=20;
set_param('MRFT_2param_optim','StopTime',num2str(simtime));
sim MRFT_2param_optim
%%%%%%%%%%%%%%%%%%%%%%%%%%%%%%%%%%%%%% %%%%%
% STEP 3 - COMPUTING IAE COST FUNCTION Q
q_opt=0;
itime=1;
for time=0:timestep:simtime
q_opt=q_opt+error(itime)*error(itime);
itime=itime+1;
end
q=q_opt;
```

8.2.2 Library of Functions for LPRS Computing

```
function J=lprsmatr(A,B,C,w)
%
% Calculation of a point of the LPRS
% of a non-integrating plant
% for matrix-vector system description,
% dx/dt=Ax+Bu; y=Cx
% w - frequency
%
n=size(A,1);
AINV=inv(A);
I=eye(n);
if w==0
J=(-0.5+j*0.25*pi)*C*AINV*B;
else
t=2.*pi/w;
AEXP=expm(0.5*A*t);
AEXP2=expm(A*t);
```

```
re_lprs=-0.5*C*(AINV+t*inv(I-AEXP2)*AEXP)*B;
im_lprs=0.25*pi*C*inv(I+AEXP)*(I-AEXP)*AINV*B;
J=re_lprs+j*im_lprs;
end

function J=lprsmatrint(A,B,C,w)
%
% Calculation of a point of the LPRS
% of an integrating plant
% for matrix-vector system description,
% dx/dt=Ax+Bu; dy/dt=Cx
% w - frequency
%
n=size(A,1);
AINV=inv(A);
AINV2=AINV*AINV;
I=eye(n);
if w==0
J=0.5*C*AINV*B-j*1000000.;
else
t=2.*pi/w;
D=expm(0.5*A*t);
re_lprs=0.25*C*AINV2*(inv(I-D*D)*(D*D-(I+2.*t*A)*D+D*D*D-I)...
+D-I)*B;
im_lprs=0.0625*pi*C*AINV*B*t+0.125*pi*C*AINV*AINV*(inv(I-D*D)...
*(3*D*D-3*D-D*D*D+I)-D+I)*B;
J=re_lprs+j*im_lprs;
end

function J=lprsmatrdel(A,B,C,tau,w)
%
% Calculation of a point of the LPRS
% for matrix-vector system description having time delay "tau",
% dx/dt=Ax+Bu(t-tau); y=Cx
% w - frequency
%
n=size(A,1);
AINV=inv(A);
I=eye(n);
if w==0
J=(-0.5+j*0.25*pi)*C*AINV*B;
else
t=2.*pi/w;
AEXP=expm(0.5*A*t);
AEXP2=expm(A*t);
```

```
AEXP3=expm(A*(0.5*t-tau));
re_lprs=-0.5*C*(AINV+t*inv(I-AEXP2)*AEXP3)*B;
im_lprs=0.25*pi*C*inv(I+AEXP)*(I+AEXP-2*AEXP3)*AINV*B;
J=re_lprs+j*im_lprs;
end


function J=lprsser200(w,name,pr)
% Function calculating the LPRS
% at a given frequency
% based on the series formula
% (as a sum of 200 terms of the series)
% 'w' - current frequency,
% 'name' - name of m-file providing
% calculation of transfer function,
% 'name' is a string variable
% 'pr' - parameters of transfer function
reloc=0;
imloc=0;
iodd=-1;
for k=1:200
iodd=-iodd;
omk=k*w;
reimloc=feval(name,omk,pr);
reloc=reloc+iodd*real(reimloc);
if iodd==1
imloc=imloc+imag(reimloc)/k;
end
end
J=reloc+j*imloc;

function J=lprs1ord(k,t,w)
%
% Calculation of a point of the LPRS
% for Transfer Function G(s)=k/(t*s+1),
% w - frequency
%
if w==0
J=k*(0.5-j*pi/4);
else
al=pi/t/w;
J=0.5*k*(1-al*csch(al)-j*0.5*pi*tanh(al/2));
end

function J=lprsint(k,w)
%
```

```
% Calculation of a point of the LPRS
% for Transfer Function G(s)=k/s,
% w - frequency
%
if w==0
J=0-j*inf;
else
J=0-j*pi*pi*k/8/w;
end

function J=lprs2ord1(k,xi,w)
%
% Calculation of a point of the LPRS
% for Transfer Function G(s)=k/(s*s+2*xi*s+1),
% w - frequency
% xi < 1
%
if w==0
J=k*(0.5-j*pi/4);
else
al=pi*xi/w;
sq=sqrt(1-xi*xi);
bt=pi*sq/w;
gm=al/bt;
b=al*cos(bt)*sinh(al)+bt*sin(bt)*cosh(al);
c=al*sin(bt)*cosh(al)-bt*cos(bt)*sinh(al);
J=0.5*k*(1-(b+gm*c)/(sin(bt)^2+sinh(al)^2))...
-j*0.25*pi*k*(sinh(al)-gm*sin(bt))/(cosh(al)+cos(bt));
end

function J=lprs2ord2(k,xi,w)
%
% Calculation of a point of the LPRS
% for Transfer Function G(s)=k*s/(s*s+2*xi*s+1),
% w - frequency
% xi < 1
%
if w==0
J=0-j*0;
else
al=pi*xi/w;
sq=sqrt(1-xi*xi);
bt=pi*sq/w;
gm=al/bt;
b=al*cos(bt)*sinh(al)+bt*sin(bt)*cosh(al);
```

```
c=al*sin(bt)*cosh(al)-bt*cos(bt)*sinh(al);
denom=sin(bt)∧2+sinh(al)∧2;
J=0.5*k*(-pi/w*sinh(al)*cos(bt)/denom+xi*(b+gm*c)/denom)...
-j*0.25*k*pi/sq*sin(bt)/(cosh(al)+cos(bt));
end



function J=lprs2ord3(k,w)
%
% Calculation of a point of the LPRS
% for Transfer Function G(s)=k*s/(s+1)∧2,
% w - frequency
%
if w==0
J=0-j*0;
else
al=pi/w;
chal=cosh(al);
shal=sinh(al);
J=k*(0.5*al*(-shal+al*chal)/shal/shal-j*0.25*pi*al/(1+chal));
end



function J=lprs2ord4(k,xi,w)
%
% Calculation of a point of the LPRS
% for Transfer Function G(s)=k*s/(s*s+2*xi*s+1),
% w - frequency
% xi > 1
%
if w==0
J=0-j*0;
else
sq=sqrt(xi*xi-1);
k1=-0.5/sq;
k2=-k1;
t1=xi+sq;
t2=xi-sq;
J=k*(lprs1ord(k1,t1,w)+lprs1ord(k2,t2,w));
end

function J=lprsfopdt(k,t,tau,w)
%
% Calculation of a point of the LPRS
% for transfer function G(s)=k*exp(-tau*s)/(t*s+1),
```

```
% 'w' - current frequency
%
if w==0
J=k*(0.5-j*pi/4);
else
al=pi/t/w;
gm=tau/t;
expal=exp(-al);
expgm=exp(gm);
J=0.5*k*(1-al*expgm*csch(al)+j*0.5*pi*(2*expal*expgm/(1+expal)-1));
end
```

Example of the LPRS computing for FOPDT dynamics.

```
script
%
% Calculation of the LPRS of FOPDT dynamics
% with ransfer function
% G(s)=k*exp(-tau*s)/(t*s+1)
% and plotting the locus
%
clear
clc
gain=1.; % gain
tconst=1.; % time constant
tdead=1.; % dead time
ommin=0.0001; % minimum frequency (in this code it is slightly higher
% than 0 to enable the use of the logarithmic scale)
ommax=1000.; % Maximum frequency
% The following code is used to generate logarithmic distribution
% of frequency points
nom=100; % number of frequency points
lmin=log10(ommin);
lmax=log10(ommax);
delta=(lmax-lmin)/(nom-1);
lom=lmin-delta;
for iom=1:nom
lom=lom+delta;
om=10^lom;
locus(iom)=lprsfopdt(gain,tconst,tdead,om);
end
plot(locus)
grid
axis('equal')
```

References

1. Aguilar, L., Boiko, I., Fridman, L., & Freidovich, L. (2012). Generating oscillations in inertia wheel pendulum via two-relay controller. *International Journal of Robust and Nonlinear Control, 22*(3), 318–330.
2. Aguilar, L., Boiko, I., Fridman, L., & Iriarte, R. (2009). Generating self-excited oscillations via two-relay controller. *IEEE Transactions on Automatic Control, 54*(2), 416–420.
3. Alexandrov, A. G. (2008). *Methods of Design of Automatic Control Systems.* Moscow: Fizmatlit (in Russian).
4. Altmann, W. (2005). *Practical Process Control for Engineers and Technicians.* Amsterdam: Newnes.
5. Åström, K. J., & Hägglund, T. (1984). Automatic tuning of simple regulators with specifications on phase and amplitude margins. *Automatica, 20,* 645–651.
6. Åström, K. J., & Hägglund, T. (1995). *PID Controllers: Theory, Design and Tuning* (2nd ed.). Durham: Research Triangle Park, NC: Instrument Society America.
7. Åström, K. J., & Hägglund, T. (2006). *Advanced PID Control.* Charlotte: Instrument Society America.
8. Atherton, D. P. (1975). *Nonlinear Control Engineering—Describing Function Analysis and Design.* Workingham: Van Nostrand.
9. Bartolini, G., Ferrara, A., & Usai, E. (1998). Chattering avoidance by second-order sliding mode control. *IEEE Transactions on Automatic Control, 43*(2), 241–246.
10. Bartolini, G., Pisano, A., Punta, E., & Usai, E. (2003). A survey of applications of second order sliding mode control to mechanical systems. *International Journal of Control, 76*(9/10), 875–892.
11. Bartolini, G., Ferrara, A., & Usai, E. (1997). Output tracking control of uncertain nonlinear second-order systems. *Automatica, 33*(12), 2203–2212.
12. Beater, P. (2007). *Pneumatic Drives: System Design, Modelling and Control.* Berlin: Springer.
13. Bequette, B. W. (2003). *Process Control—Modeling, Design, and Simulation.* New York: Prentice Hall.
14. Blickley, G. (1998). Co-creator of control loop tuning equation dies. Control Engineering. 01-01-1998. http://m.controleng.com/.
15. Boiko, I. (2009). *Discontinuous Control Systems: Frequency-Domain Analysis and Design.* Boston: Birkhäuser.
16. Boiko, I. (2005). Oscillations and transfer properties of relay servo systems—the locus of a perturbed relay system approach. *Automatica, 41,* 677–683.
17. Boiko, I. (2005). Method and apparatus for tuning a PID controller. US Patent No. 7,035,695.
18. Boiko, I., Fridman, L., & Castellanos, M. I. (2004). Analysis of second order sliding mode algorithms in the frequency domain. *IEEE Transactions on Automatic Control, 49*(6), 946–950.

19. Boiko, I. (2008). Modified relay feedback test and its use for non-parametric loop tuning. In *2008 American Control Conference*, Seattle, USA (pp. 4755–4760).

20. Boiko, I. (2008). Extension of harmonic balance principle and its application to analysis of convergence rate of second-order sliding-mode control algorithms. In *2008 American Control Conference*, Seattle, USA (pp. 4691–4696).

21. Boiko, I. (2008). Autotune identification via the locus of a perturbed relay system approach. *IEEE Transactions on Control Systems Technology, 16*(1), 182–185.

22. Boiko, I., Ernyes, A., Oli, W., & Tamayo, E. (2009). Modified relay feedback test: industrial loop tuner implementation and experiments. In *2009 American Control Conference*, Saint Louis, USA (pp. 4693–4698).

23. Boiko, I., Fridman, L., Pisano, A., & Usai, E. (2009). Analysis of input-output performance of second-order sliding mode control algorithms. *IEEE Transactions on Automatic Control, 54*(2), 399–403.

24. Boiko, I., & Sayedain, S. (2010). Analysis of dynamic nonlinearity of flow control loop through modified relay feedback test probing. *International Journal of Control, 83*(12), 2580–2587.

25. Boiko, I. (2011). Dynamic harmonic balance and its application to analysis of convergence of second-order sliding mode control algorithms. In *Proc. 2011 American Control Conference*, San Francisco, USA, 208–213.

26. Boiko, I. (2011). On frequency-domain criterion of finite-time convergence of second-order sliding mode control algorithms. *Automatica, 47*(9), 1969–1973.

27. Boiko, I. (2012). Loop tuning with specification on gain and phase margins via modified second-order sliding mode control algorithm. *International Journal of Systems Science, 43*(1), 97–104.

28. Boiko, I. (2012). Dynamic harmonic balance principle and analysis of rocking block motions. *Journal of the Franklin Institute, 349*(3), 1198–1212.

29. Castellanos, M. I., Boiko, I., & Fridman, L. (2007). Parameter identification via modified twisting algorithm. *International Journal of Control, 81*(5), 788–796.

30. Cohen, G. H., & Coon, G. A. (1953). Theoretical considerations of retarded control. *Trans. ASME, 75*, 827.

31. Corriou, J.-P. (2004). *Process Control—Theory and Applications*. Durham: ISA Press.

32. Corripio, A. B. (2001). *Tuning of Industrial Control Systems*. Durham: ISA Press.

33. Crowe, J., & Johnson, M. (1998). A phase-lock loop identified module and its application. In *IChemE Conference, Advances in Process Control*, Wales, UK.

34. Crowe, J., & Johnson, M. (1999). A new non-parametric identification procedure for online controller tuning. In *Proc. 2011 American Control Conference*, San Diego, USA (pp. 3337–3341).

35. Crowe, J., & Johnson, M. (2005). Phase-locked loop methods. In M. Johnson, & M. Moradi (Eds.), *PID Control: New Identification and Design Methods* (pp. 213–258). London: Springer.

36. Crowe, J., & Johnson, M. (2005). Phase-locked loop methods and PID control. In M. Johnson, & M. Moradi (Eds.), *PID Control: New Identification and Design Methods* (pp. 259–296). London: Springer.

37. Ellis, G. (2004). *Control System Design Guide*. Amsterdam: Elsevier.

38. Friman, M., & Waller, K. V. (1997). A two-channel relay for autotuning. *Industrial and Engineering Chemistry Research, 36*(7), 2662–2671.

39. Gelb, A., & Vander Velde, W. E. (1968). *Multiple-Input Describing Functions and Nonlinear System Design*. New York: McGraw-Hill.

40. Hägglund, T., & Åström, K. (1985). Method and apparatus in tuning a PID-regulator, US Patent No. 4549123.

41. Hang, C. C., Åström, K. J., & Wang, Q. G. (2002). Relay feedback autotuning of process controllers—a tutorial review. *Journal of Process Control, 12*, 143.

42. Hsu, J. C., & Meyer, A. U. (1968). *Modern Control Principles and Applications*. New York: McGraw-Hill.

43. Isermann, R., & Munchhof, M. (2011). *Identification of Dynamical Systems: an Introduction with Applications*. Berlin: Springer.
44. Johnson, M., & Moradi, M. (2005). *PID Control: New Identification and Design Methods*. London: Springer.
45. Kaya, I., & Atherton, D. P. (1999). A PI-PD controller design for integrating processes. In *Proc. 1999 American Control Conference*, San Diego, CA, USA (pp. 258–262).
46. Kaya, I., & Atherton, D. P. (1998). An improved parameter estimation method using limit cycle data. In *UKACC Internat. Conf. on Control, IEE* (pp. 682–687).
47. Kaya, I., & Atherton, D. P. (2001). Parameter estimation from relay autotuning with asymmetric limit cycle data. *Journal of Process Control, 11*, 429–439.
48. Krylov, N. M., & Bogolubov, N. N. (1937). *Introduction to Nonlinear Dynamics*. Kiev: Ac. Sc. of Ukrainian SSR.
49. Levant, A. (Levantovsky, L. V.) (1993). Sliding order and sliding accuracy in sliding mode control. *International Journal of Control, 58*(6), 1247–1263.
50. Lee, J., Cho, W., & Edgar, T. F. (1990). An improved technique for PID controller tuning from closed-loop tests. *AIChE Journal, 36*, 1891.
51. Ljung, L. (1999). *System Identification: Theory for the User* (2nd ed.). Upper Saddle River: Prentice Hall.
52. Lopez, A. M., Murrill, P. W., & Smith, C. L. (1967). Controller tuning relationships based on integral performance criteria. *Instrumentation Technology, 14*(11), 57.
53. Luyben, W. L. (1987). Derivation of transfer functions for highly nonlinear distillation columns. *Industrial & Engineering Chemistry Research, 26*, 2490–2495.
54. Luyben, W. L., & Luyben, M. L. (1997). *Essentials of Process Control*. New York: McGraw-Hill.
55. MacColl, L. A. (1945). *Fundamental Theory of Servomechanisms*. New York: Van Nostrand.
56. MacFarlane, A. G. J. (1979). The development of frequency-response methods in automatic control. *IEEE Transactions on Automatic Control, 24*(2), 250–265.
57. Marlin, T. E. (2000). *Process Control: Designing Processes and Control Systems for Dynamic Performance*. New York: McGraw-Hill.
58. Majhi, S., & Atherton, D. P. (1999). Autotuning and controller design for processes with small time delays. *IEE Proceedings. Control Theory and Applications, 146*(5), 415–425.
59. Majhi, S., Sahmbi, J. S., & Atherton, D. P. (2001). Relay feedback and wavelet based estimation of plant model parameters. In *Proc. 40 IEEE CDC*, Florida, USA (pp. 3326–3331).
60. Majhi, S. (2007). Relay-based identification of a class of non-minimum phase SISO processes. *IEEE Transactions on Automatic Control, 52*(1), 134–139.
61. Meirovitch, L. (2001). *Fundamentals of Vibrations*. Boston: McGraw-Hill.
62. Minorsky, N. (1922). Directional stability of automatically steered bodies. *Journal of the American Society of Naval Engineers, Inc., 34*(2), 280–309.
63. Moore, C. F., Smith, C. L., & Murrill, P. W. (1969). Simplifying digital control dynamics for controller tuning and hardware lag effects. *Instrument Practice, 23*(1), 45.
64. Najim, K., Poznyak, A., & Ikonen, E. (1996). Calculation of residence time for nonlinear systems. *International Journal of Systems Science, 27*, 661–667.
65. Najim, K., Rusnak, A., Meszaros, A., & Fikar, M. (1997). Constrained long-range predictive control based on artificial neural networks. *International Journal of Systems Science, 28*, 1211–1226.
66. O'Dwyer, A. (2006). *Handbook of PI and PID Controller Tuning Rules*. London: Imperial College Press.
67. Ogata, K. (2004). *System Dynamics* (4th ed.). New York: Prentice Hall.
68. Ogunnaike, B. A., & Ray, W. H. (1994). *Process Dynamics, Modeling and Control*. New York: Oxford University Press.
69. O'Toole, M., Bouazza-Marouf, K., & Kerr, D. (2010). Chatter suppression in sliding mode control: strategies and tuning methods. In *ROMANSY 18 Robot Design, Dynamics and Control* (pp. 109–116). Chap. 1.

70. Popov, E. P., & Paltov, I. P. (1960). *Approximate Methods of Analysis of Nonlinear Control Systems*. Moscow: Fizmatgiz (in Russian).
71. Qing-Guo, W., Chang-Chieh, H., & Qiang, B. (1999). A technique for frequency response identification from relay feedback. *IEEE Transactions on Control Systems Technology, 7*(1), 122–128.
72. Rovira, A. A. (1981). Controller algorithm tuning and adaptive gain tuning in process control. PhD Dissertation, Dept. of Chemical Engineering, Louisiana State University, Baton Rouge, USA.
73. Sayedain, S., & Boiko, I. (2011). Optimal PI tuning for flow loop, based on the modified relay feedback test. In *2011 IEEE Conference on Decision and Control*, Orlando, USA.
74. Sayedain, S. (2011). Optimal PI and PID tuning for flow loops based on the modified relay feedback test approach. MSc Thesis, University of Calgary, Calgary, Alberta, Canada.
75. Seborg, D. E., Edgar, T. F., Mellichamp, D. A., & Doyle, F. J. (2011). *Process Dynamics and Control* (2nd ed.). Hoboken: Wiley.
76. Shinskey, F. G. (1994). *Feedback Controllers for the Process Industries*. New York: McGraw-Hill.
77. Sjoberg, J., Zhang, Q., Ljung, L., Benveniste, A., Delyon, B., Glorennec, P., Hjalmarsson, H., & Juditsky, A. (1995). Nonlinear black-box modeling in system identification: a unified overview. *Automatica, 31*, 1691–1724.
78. Tan, K. K., Lee, T. H., & Wang, Q.-G. (1996). Enhanced automatic tuning procedure for process control of PI/PID controllers. *AIChE Journal, 42*, 2555–2562.
79. Tan, K. K., Wang, Q.-G., Hang, C. C., & Hägglund, T. (1999). *Advances in PID Control*. London: Springer.
80. Tsypkin, Ya. Z. (1984). *Relay Control Systems*. Cambridge: Cambridge University Press.
81. Utkin, V. (1992). *Sliding Modes in Control and Optimization*. Berlin: Springer.
82. Visioli, A. (2006). *Practical PID Control*. London: Springer.
83. Voda, A. A., & Landau, I. D. (1995). A method of autocalibration of PID controllers. *Automatica, 31*(1), 41–53.
84. Vukic, Z., Kuljaca, L., Donlagic, D., & Tesnjak, S. (2003). *Nonlinear Control Systems*. New York: Dekker.
85. Wang, Q.-G., Lee, T. H., & Lin, C. (2003). *Relay Feedback: Analysis, Identification and Control*. London: Springer.
86. Yu, C.-C. (1998). Use of saturation relay feedback in PID controller tuning, US Patent No. 5742503.
87. Yu, C.-C. (1999). *Automatic Tuning of PID Controllers: Relay Feedback Approach*. New York: Springer.
88. Ziegler, J. G., & Nichols, N. B. (1942). Optimum settings for automatic controllers. *Transactions of the American Society of Mechanical Engineers, 64*, 759–768.

Index

I. Boiko, *Non-parametric Tuning of PID Controllers*, Advances in Industrial Control,
DOI 10.1007/978-1-4471-4465-6, © Springer-Verlag London 2013